TRIZ-Anwendertag 2024

Oliver Mayer
Hrsg.

TRIZ-Anwendertag 2024

Springer Vieweg

Hrsg.
Oliver Mayer
TRIZ-Consulting
Sulzbach-Rosenberg, Deutschland

ISBN 978-3-662-70697-8 ISBN 978-3-662-70698-5 (eBook)
https://doi.org/10.1007/978-3-662-70698-5

Die Deutsche Nationalbibliothek verzeichnet diese Publikation in der Deutschen Nationalbibliografie; detaillierte bibliografische Daten sind im Internet über https://portal.dnb.de abrufbar.

Springer Vieweg ist ein Imprint der eingetragenen Gesellschaft Springer-Verlag GmbH, DE und ist ein Teil von Springer Nature.
Die Anschrift der Gesellschaft ist: Heidelberger Platz 3, 14197 Berlin, Germany

Inhaltsverzeichnis

TRIZ und die Funktionsweise unseres Gehirns

Oliver Mayer(✉)

TRIZ Consulting, Sulzbach-Rosenberg, Deutschland
o.mayer@bayern-innovativ.de

Zusammenfassung. Creativity is an important prerequisite for innovation. However, we have found that it is very difficult to be creative. And in retrospect, when you have the solution, the way to get there often seems very simple. Why is that? What prevents us from being creative? If you look at children, they are usually much more creative than adults. Why is that? In this article, various causes for the loss of creativity in elder age are presented and biological mechanisms that dampen creativity are identified. Only if you are aware of these mechanisms, you can consciously circumvent them and thus learn and train creativity again.

Schlüsselwörter: TRIZ · Gehirn · Denkmuster · Denkbegrenzungen

1 Funktionsweise des Gehirns

Von einem technischen System möchten wir üblicherweise immer sehr exakt wissen, wie es funktioniert. Je mehr Details, je mehr Zusammenhänge man kennt, desto besser. Interessanterweise ist das bei der Basis unserer menschlichen Existenz nicht der Fall. Unser Gehirn ist die zentrale Steuereinheit für alles, was wir tun, wie wir uns fühlen, wie wir agieren, wie wir reagieren etc. Aber darüber informiert man sich üblicherweise sehr wenig. Man lässt es geschehen und wundert sich dann manchmal über die eine oder andere Reaktion oder den ein oder anderen Effekt.

In diesem Text möchte ich zusammenfassen, was ich über die Funktionsweise unseres „Betriebssystems" über die Jahre durch Literatur, Erfahrungen oder durch einfache Experimente herausgefunden habe. Dies soll helfen, besser zu verstehen, wieso manche Reaktionen auf neue Ideen, auf Kreativität, auf Innovationen immer wieder in derselben Art ablaufen, obwohl es doch auch aus rein logischer Sicht anders sinnvoll wäre. Es ist kein wissenschaftlicher Text, sondern rein empirisches Wissen. Es erhebt auch keinen Anspruch auf Vollständigkeit. Es geht nicht um die medizinischen oder biologischen Grundlagen für die Funktion des Gehirns an sich (also neuronales Netzwerk, das Feuern über Axone, die Synapsen usw.) Ich versuche das übergeordnete Erscheinungsbild und die Gründe dafür darzustellen.

O. Mayer (Hrsg.): TRIZ-Anwendertag 2024,
https://doi.org/10.1007/978-3-662-70698-5_1

Der Mensch betrachtet sich gerne als ein über allen stehendes Wesen, aber im Endeffekt ist er ein biologisches System, das sich über die Evolution hin zu dem entwickelt hat, was er heute ist. Die Evolution ist kein schnelles System, sondern ein langsames, das sich kontinuierlich an neue Gegebenheiten anpasst [1]. Wenn wir uns den Zeitraum von der Aufklärung bis heute ansehen, sind das gerade einmal 250 Jahre, also sechs Generationen, in denen technisch sehr viel geschehen ist. In der Biologie ist das kein Zeitraum, nichts, in dem sich ein System adaptieren müsste. Und so kommt es, dass wir noch viele Eigenschaften haben, wie sie aus längst vergangenen Zeiten, vor 50.000 oder 100.000 Jahren, relevant waren und die uns heute noch grundsätzlich prägen, obwohl sie vielleicht sogar kontraproduktiv sind.

Vor 50.000 Jahren gab es zwar ein breiteres Angebot an Nahrung, das heißt, es war vielfältiger als heute, aber die Menge an verfügbarer Nahrung war sehr begrenzt [2]. Heute lebt die Menschheit zu einem Gutteil von drei Basisfrüchten, die wir seit 12.000 Jahren in der Landwirtschaft kultivieren: Reis, Mais und Weizen. Alle drei Basisfrüchte sind stärkehaltig. Deswegen müssen wir uns heute auch die Zähne putzen, was ja früher nicht der Fall war. Heute haben wir Nahrung im Überfluss, jedenfalls, wenn ich auf die westliche Welt schaue. Dagegen leiden immer noch 880 Mio. Menschen an extremer Mangel- und Unterernährung. Eigentlich gäbe es bei sinnvoller Umverteilung genügend Nahrung für alle. Das Problem von vor 50.000 Jahren – Mangel an Nahrung – hat sich also verändert. Statt einer breiten Palette und wenig Menge haben wir heute eine schmale Palette und große Mengen. Was hat das für Auswirkungen auf unser Gehirn? Da wenig Ressourcen zur Verfügung standen, in diesem Fall Nahrung, hat die Evolution dafür gesorgt, dass Mensch und Tier möglichst wenig Energie verbrauchen. Wie geschieht das? Der Hauptenergieverbraucher beim Menschen ist das Verdauungssystem, das ca. 40 % der Energie verwendet, um Energie aus der Nahrung zu extrahieren. Wenn Nahrung im Überfluss zur Verfügung steht, wird die Energie „gnadenlos" gespeichert. Das sieht man daran, dass in wohlhabenden Ländern meistens Übergewicht herrscht. Ich versuche meinem Bauch bewusst jeden Tag zu erklären, dass er Nahrung nicht speichern muss, weil es um die Ecke einen Supermarkt gibt. Allein, er glaubt es nicht, und ich habe keinen Einfluss darauf. So viel zu meinen bewussten Entscheidungsmöglichkeiten. Der zweitgrößte Verbraucher ist das Gehirn. Zirka 20 % der Energie werden von unserem Denkorgan benötigt. Jetzt kann man sich die Frage stellen: Wie kann das Gehirn Energie sparen, wenn es rund um die Uhr arbeitet. Die Evolution hat dafür eine Lösung gefunden, und diese Lösung heißt: Muster erzeugen. Was bedeutet das? Wenn ein Mensch etwas Neues lernen muss, dann verbraucht das Gehirn dafür sehr viel Energie. Das beobachtet man an Kindern. Sie brauchen sehr viel Energie, um das viele neue Wissen einzuspeichern. Je älter man wird, je mehr man weiß, desto mehr verfestigen sich „Muster", die dann nur noch abgearbeitet werden. Ein konkretes und für viele nachvollziehbares Beispiel ist das Autofahren. Wenn man gerade den Führerschein gemacht hat und erste Routen selbständig fährt, wird die Musik ausgeschaltet und es werden keine Diskussionen mit dem Beifahrer oder sonst wem geführt. Man konzentriert sich auf den Verkehr und die Bedienung der Technik. Man muss noch aktiv nachdenken, Kupplung drücken, Gang einlegen, in den Spiegel blicken, nach vorne schauen, lenken, Fahrrad von rechts und so weiter und sofort. Das kostet sehr viel Konzentration und damit sehr viel Energie. Nach 2 bis 3 Jahren Fahrpraxis geht das alles sehr viel flüssiger. Alle Vorgänge wurden automatisiert und in „Muster“ eingepasst. Was gleichzeitig bedeutet, dass ich als Fahrer trotz komplexer Fahrtätigkeit nebenher telefonieren, Diskussionen mit dem Beifahrer führen oder Musik hören kann. Mein Gehirn muss nicht mehr die volle Konzentration auf die Tätigkeit des Fahrens legen, sondern hat Kapazitäten frei für andere Dinge. Was das heißt? Der Ansatz, Tätigkeiten in Mustern durchzuführen, ist für das Gehirn eine Möglichkeit, Energie zu sparen.

Diese Möglichkeit hat aber auch eine Kehrseite. Wenn ein neues Problem auftaucht, etwas Unbekanntes, etwas, wo man kreativ sein muss, versucht das Gehirn als erstes, dieses Problem mit bestehenden Mustern zu lösen und nicht unbedingt auf das Neue einzugehen. Wenn sich abzeichnet, das ist doch etwas Neues und hier muss etwas getan werden, kommen automatisch alle Argumente hoch, die gegen das Neue sprechen. Man kennt das aus alltäglichen Situationen. In einer Besprechung wird ein neues Projekt oder eine neue Herausforderung vorgestellt. Dann erfolgt nicht ein „Hurra, endlich was Neues, wir können kreativ sein“, sondern eher ein „Da spricht doch so viel dagegen, das ist so kompliziert“ und so weiter. Es werden eher die bremsenden Argumente vorgebracht. Neues bedeutet eigentlich Gedankenarbeit und das ist anstrengend. „Ich habe doch sowieso schon genügend zu tun, warum das jetzt auch noch“, usw. Diese Situation kann man nur durch Bewusstsein übersteuern. Aber auch hier gibt es noch ein biologisches Hemmnis.

Unser Gehirn denkt assoziativ [3]. Es nimmt über seine Sinnesorgane Informationen auf und sucht mit unglaublicher Geschwindigkeit sofort nach eigenen Erfahrungen und Wissen, um es mit diesen Informationen abzugleichen bzw. zu korrigieren, sei es in positiver Hinsicht mit Zustimmung oder in negativer Hinsicht: „Nee, da stimmt was nicht, das kenne ich anders“. Wir arbeiten also bevorzugt auf „Triggerbasis“. Wir bekommen einen Impuls und darauf reagiert unser Gehirn und liefert Informationen, die dazu passen. Und diese Informationen wollen dann auch kommuniziert werden. Vielleicht kennt man das noch aus der Schule. Der Lehrer fragt etwas. Ich kenne die Antwort und melde mich. Und ich werde ignoriert. Dann baut sich ganz schnell Ärger auf, weil man nicht drangekommen ist und sein Wissen nicht mitteilen konnte. Das Wissen muss raus. Aus diesem Grund sind auch Brainstormings nicht ganz einfach, weil üblicherweise in bestehenden Runden häufig dieselben Leute als erstes einen Gedanken formulieren und alle anderen dann automatisch auf diesem Gedanken aufbauen. Es ist nicht ganz einfach, von dieser Basis aus neue Ideen oder Ansätze durchzubringen.

Und hier sind wir wieder bei den Mustern. Wenn ich ein neues Projekt, eine neue Aufgabe vorstelle, kommen, wie berichtet, üblicherweise zuerst eher negative oder hindernde Argumente, die kommuniziert werden wollen. Das kann ich nicht unterdrücken. Deshalb ist hier eine gute Strategie, die spontane Kommunikation zuzulassen, jedoch in dem Bewusstsein, dass vornehmlich Widerstand gegen Neues kommt. Um dann zu sagen: „Okay, jetzt, nachdem alles gesagt ist, können wir richtig anfangen, das Problem zu analysieren“. Die Fähigkeit, dies zuzulassen ist eine schwierige Aufgabe: Unser Gehirn ist in der Lage, fünf bis sieben Informationseinheiten parallel zu verarbeiten. Eine höhere Anzahl von zu verarbeitenden Einheiten bedeutet gedanklichen Mehraufwand, und schnell kommt das Gefühl hoch: Dann lassen wir es doch lieber sein. Zur Resignation führt daneben auch die Vielzahl der Gegenargumente und damit der zu verarbeitenden Informationseinheiten. Dann sagt man üblicherweise: „Tschuldigung, war ja nur `ne Idee“ und lässt sie beiseite. Hier ist es möglich, mit Bewusstsein gegenzusteuern und zu entgegnen: „Die Gegenargumente sind berechtigt und wir werden trotzdem daran arbeiten.“ Dies muss aber immer bewusst geschehen.

Eine dritte Eigenart unseres Gehirns kommt auch aus früheren Zeiten von vor 50.000 Jahren. Damals war das Leben unsicherer als heute. Man war in der Steppe, Prärie oder sonst irgendwo unterwegs und immer von Tieren bedroht. Das hatte zur Folge, dass die Evolution uns Menschen genauso wie den Tieren ein Grundprogramm eingegeben hat, das aus drei einfachen Regeln besteht. Wenn ich etwas sehe, werden schlagartig folgende Entscheidungen getroffen:

1. Das ist etwas zum Fressen. Das jage ich, da geh ich drauf los.
2. Das ist etwas zum Fürchten. Ich renne weg.
3. Das ist neutral. Das ignoriere ich.

Diese Entscheidungen müssen heute natürlich nicht mehr bezüglich gefährlicher Tiere getroffen werden. Was aber geblieben ist, ist die Tatsache, dass man im Falle neuer Herausforderungen, sofort nach einer Lösung für diese Situation sucht. Und diese Lösung muss dann auch kommuniziert (siehe oben) oder ausgeführt werden.

Das bedeutet bezüglich der Problemstellung in einem Brainstorming, dass automatisch über die Assoziation nach Lösungen dafür gesucht wird und diese Lösungen auch kommuniziert werden müssen. Als Moderator sollte man nicht sagen: „Leute, das ist das Problem und jetzt gehen wir das mal ganz strukturiert an und wollen es richtig verstehen, bevor wir an Lösungen denken.“ Das funktioniert nicht, denn unser Gehirn wird automatisch assoziative Lösungsvorschläge „ausspucken“ wollen, die kommuniziert werden müssen. Das bedeutet für eine neue Situation: Treffen in einem kurzen 20-minütige Brainstorming, in dem alle zu der Situation sagen dürfen, was ihnen einfällt. Wenn dabei Lösungsideen zum Vorschein kommen, notiert man sie am besten mit den Namen, der Personen, die sie aufgebracht haben. Nach dem Sammeln erster Reaktionen beginnt man, das Problem tiefgründig zu analysieren und die Lösungsvorschläge zu bearbeiten. Notierte und namentlich zugewiesene Lösungsideen aus dem ersten Meeting können bei Weiterbearbeitung auch später noch den kreativen Köpfen zugeordnet werden. Dies bringt Anerkennung für die Person, die die Idee schon zu Anfang kommuniziert hat.

2 Zusammenhang mit TRIZ

Was bedeutet die Ausführung nun für die Kreativität, um Innovationsideen oder Lösungsideen zu schaffen? Es wurde dargestellt, wie die Grundfunktionalität unseres Gehirns die Kreativität beeinflusst. Genau an diesem Punkt setzt die TRIZ-Methode ein. Diese Methode bietet einen Ansatz und Ablauf, wie man diese Nachteile der Funktionsweise unseres Gehirns umgehen beziehungsweise kompensieren kann. Hat man beispielsweise ein neues Problem, wird man durch die Anwendung von TRIZ dazu angeleitet, nicht die negativen Gefühle von „Das ist schwierig, das brauche ich nicht, das will ich nicht“ die Oberhand behalten zu lassen, sondern man wird aktiv hingeführt, das Problem bezüglich Verständnisses, Ressourcen, Wissensmangel, Funktionsweise etc. zu analysieren, um damit über das Bewusstsein die grundlegende Ablehnung zu kompensieren. Ein weiterer Punkt ist die TRIZ-Frage, ob das, was als Idee generiert wurde, die einzige Idee ist und auch, ob sie überhaupt die beste Lösung darstellt.

Die Innovationsprinzipien nutzen die assoziative Arbeitsweise unseres Gehirns, indem sie Trigger vorgeben, über die man nachdenkt. Aufgrund der assoziativen Arbeitsweise liefert das Gehirn Lösungsideen, ohne selbst danach suchen zu müssen [4].

Die Aufteilung eines großen Problems in Miniprobleme, die oftmals über Widersprüche strukturiert werden, erleichtert es unserem Gehirn ebenfalls, Lösungen zu finden. Je kleinteiliger ein Problem ist, desto einfacher funktioniert das assoziative Denken. Generelle und große Probleme, wie zum Beispiel „Wie rette ich die Welt vor dem Klimawandel"? sind für unser Gehirn zu komplex, um über Assoziationen zu einer Lösung zu gelangen.

3 Zusammenfassung

Es gibt noch weitere Eigenschaften unseres Gehirns. Wichtig ist dabei, dass man diese nicht einfach abschalten kann, sondern, dass man sie akzeptieren muss und am besten bewusst zulässt, um dann in einen Zustand zu kommen, mit dem man Probleme systematisch, zum Beispiel mit TRIZ, lösen kann. Insofern ist es sinnvoll, sich generell mit der Funktionsweise unseres Gehirns auseinander zu setzen und zu verstehen, welche Mechanismen wirken, um entsprechende Kreativitätswerkzeuge noch besser einsetzen zu können.

Literatur

1. Kahneman, D.: Schnelles Denken, langsames Denken, 16. Aufl. Penguin (2011). ISBN 978-3-328-10034-8
2. Harrari, Y.N.: Eine kurze Geschichte der Menschheit, 37. Aufl. Pantheon (2013). ISBN 978-3-570-55269-8
3. Spitzer, M., Herschkowitz, N.: Wie Kinder denken lernen, 3. Aufl. mvgverlag-Beckmann (2022). ISBN 978-3-7474-0002-9
4. Spitzer, M.: Digitale Demenz. Droemer (2012). ISBN 978-3-426-30056-5

AI and TRIZ: Improved Search for a Leading Area within Function Oriented Search

Jochen Wessner(✉)

Esslingen, Deutschland
jochen.wessner2@de.bosch.com

Abstract. The paper tries to answer the question if AI tools like chatGPT could help to improve the more or less subjective search for leading areas or industries within Function Oriented Search. The obtained results imply that AI tools could support the search in an effective and objective way, especially if prompt templates with adjusted user inputs for the specific search are applied. Furthermore, the search for the leading area should be extended to nature. A Nature Oriented Innovation Strategy was introduced by Hill and this work could be extended to prompts for a Function Oriented Search in Nature (FOSN). Nevertheless, the human factor still remains because the creative work with the problem, i.e., deriving the core problem, prompt creating and decision making seems to be difficult to hand over to an AI completely. Especially, the definition of the generalised function, its parameters and the goal to be achieved are core issues carried out by individuals with differing skill.

Keywords: AI · Function Oriented Search · Leading Area · chatGPT

1 Introduction

The current version of Function Oriented Search was proposed by Litvin in 2004 [1]. This TRIZ problem solving tool is based on the fact that a similar problem formulated on a general functional level was already solved elsewhere because in the other area the solving was vitally important for the system to work. This elsewhere is the leading area and unfortunately the systematic determination of the leading area is only roughly described. In [2, 3] first trials with an image and key word search were made using already present internet resources.

The public availability of Large Language Model AI tools like bing Copilot (Microsoft) or ChatGPT (openAI) offers additional support to find possible leading areas. The use of these tools to support the innovation process in general and the developing of new ideas is described by various authors [4–8]. Now, this paper tries to answer the question if those tools could help to reduce the more or less subjective search for leading areas within Function Oriented Search.

O. Mayer (Hrsg.): TRIZ-Anwendertag 2024,
https://doi.org/10.1007/978-3-662-70698-5_2

2 AI Tools and Function Oriented Search

The algorithm of Function Oriented Search defined by Litvin has 8 steps (see Fig. 1). Carrying out the steps up to step 4 is quite straightforward. The determination of the leading area or industry in which the similar function is vitally important is s step which is not clearly stated.

Nowadays a great variety of AI tools are publicly available. The author believes that these tools could improve the search for a leading area and reduce the level of subjectiveness to define it.

After the completion of step 4 (see Fig. 1) of Function Oriented Search the necessary parameters and a generalised function are defined. Now, the objective is to create

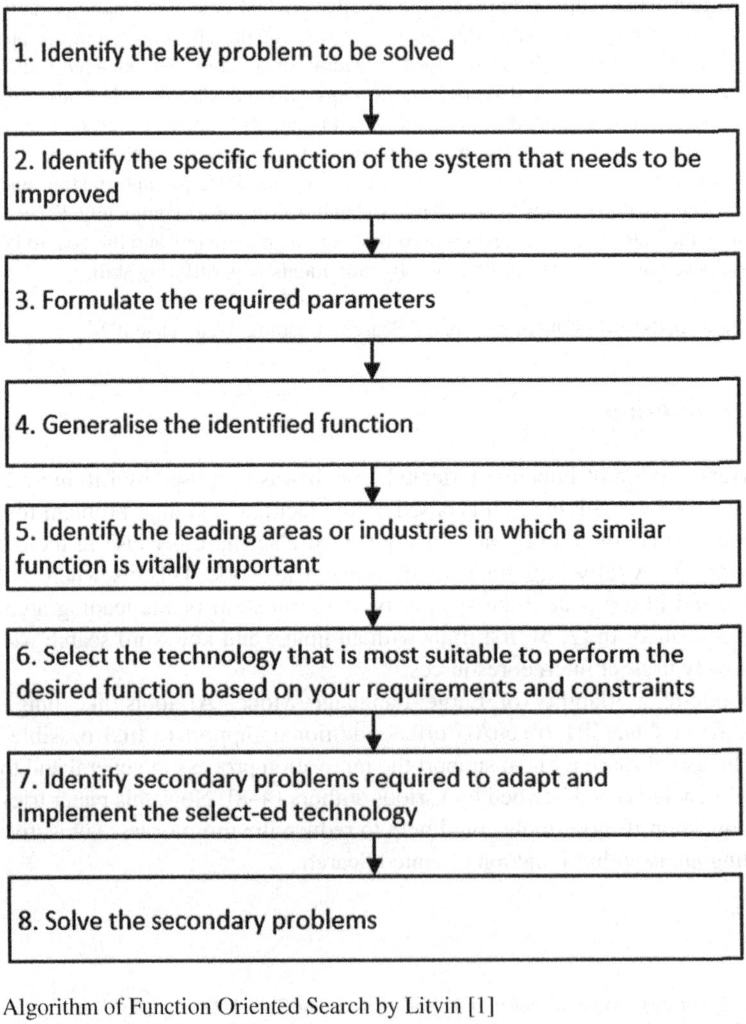

Fig. 1. Algorithm of Function Oriented Search by Litvin [1]

a text-based AI prompt to determine a leading area in which this generalised function is vitally important for the function fulfilment of a system. Ideally, first solution ideas are discovered as well. A subsequent second search could be created and carried out after the initial search within this leading area to come up with additional details to support the implementation.

The workflow is supposed to be particularly helpful in the case of great uncertainty where to look for solutions. Unfortunately, the setting up of a suitable prompt is not straightforward and the influence of the human factor is transferred to prompting and decision making what area to select. Therefore, it seems unlikely that the process could be completely automated. Personal skill and knowledge are still an important factor to develop a good concept, usually from a combination of different ideas. But not only prompt development might pose a problem, changing AI tools, license issues and the changing information technology and its environment could create difficulties which are hard to foresee.

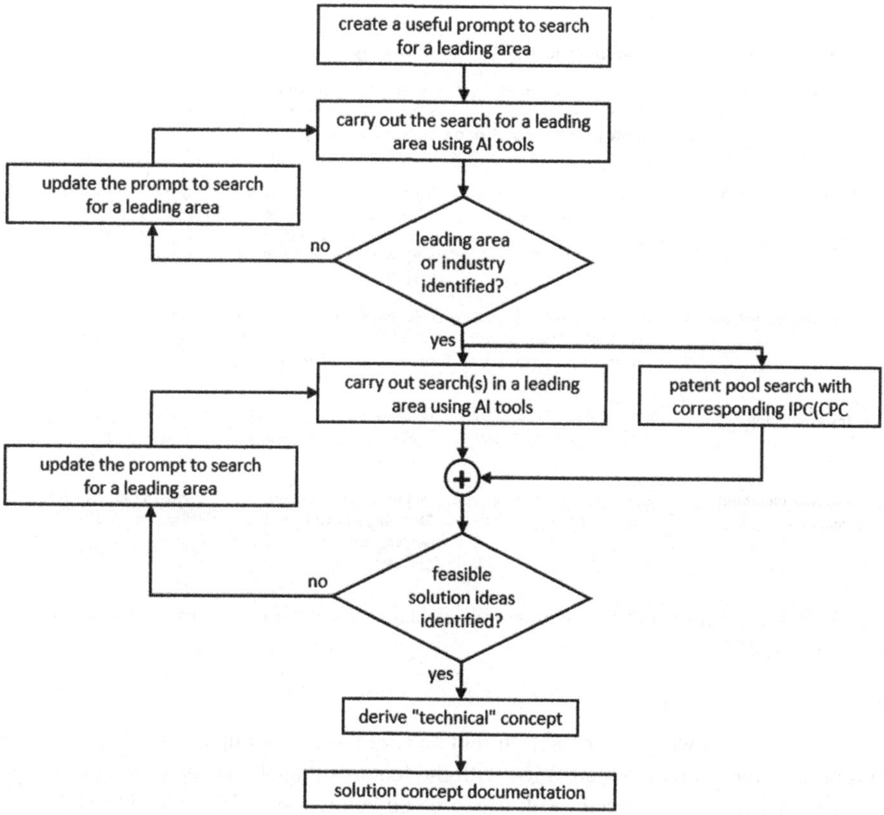

Fig. 2. Workflow for identifying a leading area and derive a solution concept

The success of the search is further influenced by the quality of the analysis of the initial situation. The definition of the core problem is very important, i.e., the right parameters and the right generalised function have to be identified, to find a satisfying result and a good solution.

The issue of prompt creation was also addressed by Tanasak and Adunka [9]. An example for Function Oriented Search with context and instructions is given below:

[Context] In filter industry a filter element has to be fitted into a filter casing. The goal is to allow only original filter elements to be assembled and hinder other filter elements from being assembled. [Instruction] Help find other industries that are not related to filter industry. Find 5 more industries which perform similar function, and briefly explained in tabulated format including (Industrial Domain, Function Similarity, Technology, Action Principle).

Figure 3 shows the result using the AI tool Bing Chat (GPT-4, 2024) and the prompt above.

Tabelle

Industrial Domain	Function Similarity	Technology	Action Principle
1. Automotive Manufacturing	Ensuring proper assembly of vehicle components (e.g., engine parts, chassis, electrical systems)	**Robotic Assembly Lines:** Robots precisely fit and secure components during vehicle assembly.	Components are aligned, bolted, welded, or glued together.
2. Electronics Assembly	Ensuring correct placement of electronic components on circuit boards	**Surface Mount Technology (SMT):** Automated machines place tiny components (resistors, capacitors, ICs) onto PCBs.	Components are soldered onto the board.
3. Furniture Manufacturing	Properly assembling furniture pieces (e.g., chairs, tables, cabinets)	**CNC Machining and Joinery:** Computer-controlled machines cut, shape, and join wood pieces.	Joints (dovetail, mortise and tenon) securely connect parts.
4. Toy Production	Assembling toy components (plastic, metal, fabric) into finished toys	**Injection Molding:** Molten plastic is injected into molds to create toy parts.	Parts are snapped, glued, or screwed together.
5. Pipe and Plumbing Systems	Connecting pipes, valves, and fittings for fluid transport	**Welding, Soldering, and Threaded Joints:** Pipes are welded, soldered, or threaded to create a sealed system.	Components are joined to form a continuous pipeline.

Fig. 3. Result for a prompt by Tanasak and Adunka to assemble only correct filter parts (Bing Chat, GPT-4, 2024)

To get an idea what other issues might be addressed to set up a suitable prompt the author used the prompt "create a list with the features that should be present in a good chatgpt prompt". The yielded result is expressed below (chatGPT3.5 Turbo, 2024):

- Clear and specific instructions: A good chatGPT prompt should provide clear and specific instructions to guide the model's response. This helps in obtaining accurate and relevant answers.

- Contextual information: Including relevant background information or context in the prompt can help the model understand the desired response better. This can include details about the topic, setting, or any specific constraints.
- Desired format or structure: If there is a specific format or structure expected in the response, it should be clearly mentioned in the prompt. For example, if the response needs to be in bullet points or paragraphs, it should be specified.
- Examples or sample responses: Providing examples or sample responses can help the model understand the expected output. This can be particularly useful when dealing with subjective or opinion-based questions.
- Limitations or constraints: If there are any limitations or constraints on the response, such as word count limits or specific sources to refer to, they should be clearly stated in the prompt.
- Tone or style guidelines: If there are any specific tone or style guidelines to be followed in the response, they should be mentioned. For in-stance, if the response needs to be formal or casual, it should be specified.
- Error handling instructions: In case the model encounters an error or is unable to generate a response, it is helpful to include instructions on how to handle such situations. This can include suggestions for alternative responses or actions to take.
- Ethical considerations: If there are any ethical considerations or guidelines to be followed while generating the response, they should be clearly outlined in the prompt. This can include avoiding biased or dis-criminatory content.
- User engagement instructions: If the prompt requires the model to engage the user in a specific way, such as asking follow-up questions or seeking clarification, it should be mentioned explicitly.
- Feedback or evaluation instructions: If the prompt is part of a feed-back or evaluation process, instructions on how to provide feedback or evaluate the response should be included. This can help in improving the model's performance over time.

The list provides additional aspects which should be kept in mind. Especially feedback and stepwise prompting could lead to good results in a desired direction. As mentioned earlier, the aim of the method is to increase the probability to find a leading area and support the optional search within the leading in a in less subjective way. The need of iterations of both prompts is very likely and outlined in Fig. 2. If a suitable leading area is found the iteration process could be stopped early and it is possible to return to Function Oriented Search proposed by Litvin continuing with step 6 (see Fig. 1). A variant of the proposed approach would be to search the patent pool using the specific search engines of the various databases. Ideally, many solution ideas are retrieved which could finally combined to a feasible solution concept (Fig. 3).

The Function Oriented Search prompt template is limited to this specific TRIZ tool. Nevertheless, the idea of AI support and prompt engineering for other TRIZ tools is currently thought of by Tanasak and Adunka [9].

Below two examples are shown to illustrate the process in more details. The first example features a tea press in which the stability of the long, thin rod should be improved. The second example features toy building bricks which should be changed to improve the stability of a created building block structure (see Fig. 4). In [2, 3] the problems are described in more detail.

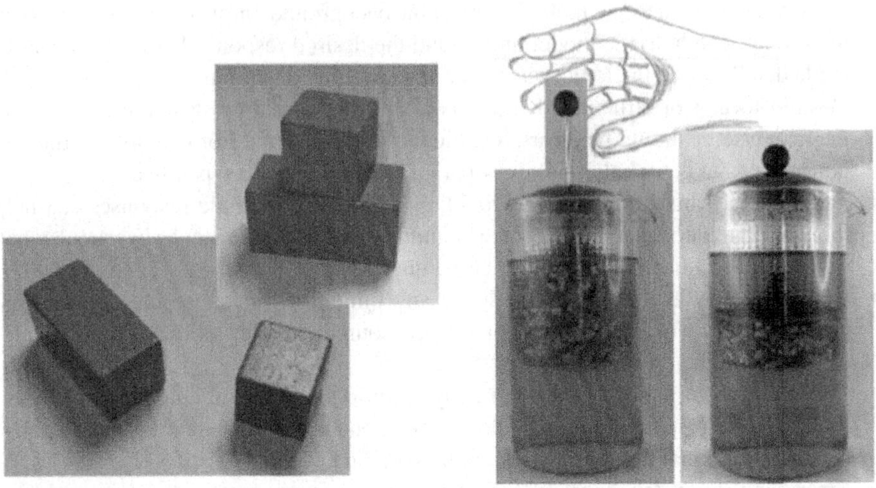

Fig. 4. Wooden toy building bricks (left) and tea press (right)

Area	Functional Similarity	Technology	Action Principle
Coffee Brewing	Uses a plunger to separate grounds from the liquid	French Press	The coffee grounds are steeped in hot water, then separated by pressing down the plunger
Juice Extraction	Uses pressure to extract juice from fruits	Manual Citrus Juicer	The fruit is pressed down onto a cone, separating the juice from the pulp
Oil Extraction	Uses pressure to extract oil from seeds	Oil Press Machine	Seeds are placed in a container and pressure is applied to extract the oil
Wine Making	Uses pressure to extract juice from grapes	Wine Press	Grapes are placed in the press and pressure is applied to extract the juice
Cheese Making	Uses pressure to form cheese from curds	Cheese Press	Curds are placed in a mold and pressure is applied to form the cheese

Fig. 5. Result for a prompt to improve the tea press (Bing Chat, GPT-4, 2024)

The initial tea press prompt looks like this:

[Context] In the home appliance sector a tea press is used to make tea. The tea press has a body to hold the hot water and a plunger which is pressed down with a rod after seeping. The goal is to avoid bending of the rod. [Instruction] Help finding at least 5 areas or sectors which perform a similar function and briefly explain in a table format the area, the functional similarity, the technology and the action principle.

Figure 5 shows the obtained results.

The indicated solution areas look suitable, but the value of the results is thought to be poor. The use of press and pressing might have created an unsuitable priming. Therefore, the prompt was changed to:

[Context] Long and thin parts have the tendency (to buckle or) to bend. The goal is to avoid bending (or buckling) of a long and thin object. [Instruction] Help finding at least 5 areas or sectors which perform a similar function and briefly explain in a table format the area, the functional similarity, the technology and the action principle.

Area	Functional Similarity	Technology	Action Principle
Construction	Steel beams used in construction are long and thin, yet they resist bending due to	Used in the construction of buildings and bridges.	The I-beam structure provides strength and rigidity, preventing bending.
Aerospace	The wings of an airplane are long and thin, yet they resist bending due to their	Used in the design of aircraft.	The internal structure of the wings, often a honeycomb structure, provides strength and
Automotive	The chassis of a car is a long and thin structure that resists bending.	Used in the design of vehicles.	The chassis is designed to distribute stress evenly, preventing bending and deformation.
Marine	The masts of ships are long and thin, yet they resist bending due to their circular cross-section and the use of guy-wires.	Used in the design of ships.	The circular cross-section and the use of guy-wires distribute stress and prevent bending.
Medical	Medical devices such as catheters and endoscopes are long and thin, yet they resist bending due to their flexible yet	Used in various medical procedures.	These devices are designed to be both flexible and rigid, allowing them to navigate through the body without bending or breaking.

Fig. 6. Result for an iterated prompt to improve the tea press (Bing Chat, GPT-4, 2024)

Figure 6 shows the result for the changed initial prompt.

The use of the parameters long and thin improved the search significantly.

To extend the search to nature the following prompt was used:

[Context] Long and thin parts have the tendency to bend. The goal is to avoid bending of a long and thin object. [Instructions] Help finding at least 7 ways in flora and fauna (nature) where a similar function is performed and briefly explain in a table format the area, the functional similarity, the technology and the action principle.

Figure 7 states the result for the usage of "flora and fauna" and represents a first iteration because the application of "nature" returned unsatisfactory answers.

Leaving the tea press and moving on to the toy building bricks, the initial prompt

Area	Functional Similarity	Technology	Action Principle
Bamboo	Bamboo stalks are long and thin, but resist bending due to their hollow, segmented structure	Plant structure	The hollow, segmented structure of bamboo provides strength and flexibility
Bird Feathers	Feathers are long and thin, but resist bending due to their structure	Animal structure	The structure of feathers, with a central shaft and branching barbs, provides strength
Elephant Trunks	Elephant trunks are long and thin, but resist bending due to their muscular structure	Animal structure	The muscular structure of the trunk provides strength and flexibility
Spider Silk	Spider silk is thin but extremely strong and resistant to bending	Animal structure	The molecular structure of spider silk gives it high tensile strength
Cacti	Cacti have long, thin structures (spines) that resist bending	Plant structure	The hard, fibrous structure of cacti spines provides strength
Pine Needles	Pine needles are long and thin, but resist bending due to their structure	Plant structure	The structure of pine needles, with a thick cuticle and sunken stomata, provides strength
Grass	Grass blades are long and thin, but resist bending due to their structure	Plant structure	The structure of grass blades, with a central vein and parallel venation, provides strength

Fig. 7. Result for a prompt to improve the tea press using ideas from flora and fauna (Bing Chat, GPT-4, 2024)

looks like the following:

[Context] In the toy sector wooden building blocks are used to create buildings. The goal is to increase the stability of the created structures during playing. [Instruction] Help finding at least 5 areas or sectors which perform a similar function and briefly explain in a table format the area, the functional similarity, the technology and the action principle.

Area	Functional Similarity	Technology	Action Principle
Civil Engineering	Building stable structures	Use of materials like steel, concrete	Physics and engineering principles to ensure stability
Architecture	Designing buildings	CAD software, physical models	Balancing aesthetics and functionality while ensuring stability
Automotive Industry	Assembling parts to create a stable vehicle	Robotics, assembly lines	Precise assembly to ensure vehicle stability and safety
Software Development	Building stable software systems	Programming languages, development tools	Modular design and testing to ensure software stability
Molecular Biology	Creating stable molecular structures	Bioinformatics tools, laboratory equipment	Understanding molecular interactions to build stable structures

Fig. 8. Result for the initial prompt to improve the building bricks (Bing Chat, GPT-4, 2024)

Figure 8 presents the result which show a reference to the prompt but overall it is not very useful. Therefore, an iteration is carried out by exchanging stability with "sticking together" and by adding the parameter "non-permanent". This stresses the temporal limitation of the created structure.

[Context] In the toy sector wooden building blocks are used to create buildings. The goal is to increase the non-permanent sticking together of the building blocks. [Instruction] Help finding at least 5 areas or sectors which perform a similar function and briefly explain in a table format the area, the functional.

Result in Fig. 9.

Area	Functional Similarity	Technology	Action Principle
Modular Construction	Building temporary structures	Prefabricated modules	Modules can be joined and separated as needed
Theatre Set Design	Creating temporary sets	Various materials, including wood and paint	Sets are designed to be assembled and disassembled quickly
Exhibition Stands	Building temporary displays	Modular systems, graphics	Stands are designed for easy assembly and disassembly
Toy Industry (Lego)	Building models	Interlocking plastic bricks	Bricks can be joined and separated easily
Magnetic Toys	Creating temporary structures	Magnetic pieces	Magnetic attraction allows pieces to stick together and be separated easily

Fig. 9. Prompt iteration to improve the building bricks (Bing Chat, GPT-4, 2024)

The prompt above also yields some first solution ideas. Prompting with nature, i.e., with flora and fauna gives:

[Context] In the toy sector wooden building blocks are used to create buildings. The goal is to increase the non-permanent sticking together of the building blocks during playing. [Instructions] Help finding at least 7 ways in flora and fauna where a similar function is performed and briefly explain in a table format the area, the functional similarity, the technology and the action principle.

Result in Fig. 10.

A prompt for a refined search in identified areas could look like this:

Area	Functional Similarity	Technology	Action Principle
Velcro	Temporary adhesion	Hook and loop fasteners	The hooks of one side stick to the loops of the other, creating a bond that can be easily undone.
Gecko feet	Temporary adhesion	Van der Waals forces	The tiny hairs on a gecko's feet exploit these forces to stick to surfaces and can be easily detached.
Barnacles	Temporary adhesion	Cement-like substance	Barnacles secrete a substance that hardens, allowing them to stick to surfaces. They can be removed without damaging the surface.
Spider webs	Temporary adhesion	Sticky silk	Spiders spin sticky silk that can catch prey but can be easily cleaned or removed.
Clinging plants (e.g., ivy)	Temporary adhesion	Aerial rootlets	These plants have small roots that can cling to surfaces but can be removed without causing damage.
Mussel adhesive proteins	Temporary adhesion	Protein-based adhesive	Mussels secrete proteins that allow them to stick to surfaces underwater. The bond can be broken without causing damage.
Burdock seeds	Temporary adhesion	Hooked barbs	The hooked barbs of burdock seeds allow them to stick to fur or clothing and can be easily removed.

Fig. 10. Result for a prompt to improve the building bricks using flora and fauna in the prompt (Bing Chat, GPT-4, 2024)

[Context] In flora and fauna velcro, gecko feet and burdock seeds could help to create stable but non-permanent structures. [Instruction] Explain in a table format and in detail the functional principle, list at least 5 examples in technology where and how it is applied and list image links (Fig. 11).

Flora/Fauna	Functional Principle	Examples in Technology	Image Links
Velcro	Velcro was inspired by the burdock plant. The seeds of these plants disperse by attaching to any animal that happens to rub against it. The seeds have hundreds of little hooks on their surface that can	1. Used in clothing and footwear; 2. Used in diapers; 3. Used in automobiles; 4. Used in space shuttles; 5. Used in surgery	Image Link
Gecko Feet	Geckos have many microscopic hairs, or setae, on their feet that increase the Van der Waals forces - the distance-dependent attraction between atoms or molecules - between its feet and the surface.	1. Dry adhesive inspired by gecko feet; 2. Sticky handheld paddles for soldiers; 3. Robots able to pull objects hundreds of times their own weight; 4. Grippers for space repairs; 5. LEMUR robot for inspecting and maintaining installations on the International Space Station.	Image Link
Burdock Seeds	The burdock has long hooked seeds — similar to miniature walking sticks — shaped in such a way to help them spread by catching on to passing creatures.	1. Velcro; 2. Biorefining burdock seeds and roots; 3. Production of fine chemicals and value-added compounds; 4. Synthesis of 5-Hydroxymethylfurural (5-HMF); 5. Synthesis of sulfonated magnetic iron-carbon catalysts.	Image Link

Fig. 11. Refined search in leading areas (example building bricks, Bing Chat, GPT-4, 2024)

A prompt to identify patents, claims or IPC numbers was not successful. The answer only suggested to use specific patent database search engines.

3 Results

The obtained results for the two examples above show that AI tools could be used to find a leading area within Function Oriented Search. Nevertheless, the quality of the prompt defines the quality of the results. This is up to the skill of the prompt creator and additional human factors which are still present (see list below):

- analysing the problem situation (derive the adequate generalised function and the important parameters)
- creating a suitable/useful prompt (optional: AI supported)
- refining of the prompt
- evaluating the results
- making the decision what leading area(s) are chosen
- searching the leading area (optional)
- choosing suitable solution ideas
- creating a solution concept
- criticising the solution concept (looking for secondary problems)
- improving the solution concept (solving the secondary problems)
- implementing the solution concept

For creating a good prompt the following items should be taken into account (see tea press example):

- stating the formulation of the goal (*improve stability* of the rod) and/or of the inverted goal (*avoid bending and/or buckling* of the rod)
- using parameters (*long and thin* rod) or refraining from using parameters (rod)
- creating a prompt template matrix with the two variants above (mixing the formulation of the goal or of the inverted goal with using or not using parameters)
- priming or creating context (attention: the multiple use of the terms "press" and "pressing" resulted in a wrong solution direction)
- clear and concise instructions (use the 5 W's to be precise: What? Where? When? Why? For Whom?)
- anchoring (for second searches within a certain area)

With a suitable initial prompt and some iterations, a good quality was achieved in a pretty short time for both examples which implies that the application of AI tools increases the efficiency of the search for a leading area within Function Oriented Search. The available tools change rapidly. Also, a tendency is noticed that the different applications and providers try to foster charged services by reducing the number of free accesses. This hinders stepwise prompting significantly. A critical issue is the use of sensible or classified data. Analysing the requests might reveal possible company strategies or development directions. This makes the use within a company difficult or precludes it completely.

The trial to add a search for patent information, e.g., patent numbers or claim references, to the prompt was not successful. Instead of direct results the AI search tools proposed the search within the available patent databases.

In addition, only already known and taught elements could be found by AI and in most cases the whole process resembles very much a black box procedure. Nevertheless, the obtained results imply that the use is effective. A lot of ideas and starting points were obtained in a very short time.

The human factor for setting up prompts and decision making was already hinted. Prompt templates, ideally stored within a database might help to reduce subjectivity or the creation of unintended bias. As for decision making, evaluation methods and rating tools are available to make this more objective.

4 Summary and Conclusions

The aim of the paper is to answer the question if AI tools like bing Copilot with GPT-4 could help to reduce the more or less subjective search for leading areas or industries within Function Oriented Search.

The obtained results show that AI tools could support the search in an effective and objective way, especially if prompt templates are created which are only adjusted by a user in a minimal way. Currently, Robert Adunka sets up the project ccTOPP with the aim of creating useful and suitable prompts for different areas within TRIZ. The results also showed that the search for the leading area should be extended to nature. A Nature Oriented Innovation Strategy was introduced by Hill and this work could be extended to prompts for a Function Oriented Search in Nature (FOSN).

Nevertheless, the human factor still remains because the creative work with the problem (deriving the core problem, prompt creating and decision making) seems to be difficult to hand over to an AI completely.

Here, the 5W's could help to work with the problem in a creative way and set up a suitable prompt after definition of the generalised function, its parameters and the goal to be achieved.

References

1. Litvin, S.: New TRIZ-based tool – function-oriented search (FOS). In: Cascini, G. (eds.) TRIZ Future Conference 2004. Firenze University Press (2004). ISBN 88-8453-221-3
2. Wessner, J.: The combined use of publicly available dictionaries and search engines to find solution ideas. In: Mayer, O. (eds.) TRIZ-Anwendertag 2022, pp. 23–33. Springer Vieweg, Berlin (2022). https://doi.org/10.1007/978-3-662-66201-4_4
3. Wessner, J.: The use of publicly available image search engines to find solution ideas efficient use of TRIZ information resources. In: Nowak, R., Chrząszcz, J., Brad, S. (eds.) Systematic Innovation Partnerships with Artificial Intelligence and Information Technology. TFC 2022. IFIP Advances in Information and Communication Technology, vol. 655, pp. 109–120. Springer, Cham (2022). https://doi.org/10.1007/978-3-031-17288-5_10
4. Agulyansky, A.: Identifying customer needs: harnessing AI for effective problem solving and innovation. Workshop Presentation ETRIA World Conference TRIZ Future 2023 – TFC2023 (12–14 September 2023). Offenburg University
5. King, R.: Chaotic Problems, TRIZ, and AI-Universal Problem Solving. Workshop Presentation ETRIA World Conference TRIZ Future 2023 – TFC2023 (12–14 September 2023). Offenburg University
6. Tsourikov, V.: AI as a Dominant Innovator. Workshop Presentation ETRIA World Conference TRIZ Future 2023 – TFC2023 (12–14 September 2023). Offenburg University
7. Mitchell, I.: AI and its Possible Impact. Workshop Presentation ETRIA World Conference TRIZ Future 2023 – TFC2023 (12–14 September 2023). Offenburg University
8. Livotov, P.: AI-assisted Invention with ChatGPT and Automatic Idea Generator. Workshop Presentation ETRIA World Conference TRIZ Future 2023 – TFC2023 (12–14 September 2023). Offenburg University
9. Tanasak, P., Adunka, R.: TRIZ and Generative AI – V3.0. ResearchGate (2024). https://doi.org/10.13140/RG.2.2.17134.42563. https://www.researchgate.net/publication/380036640_TRIZ_and_Generative_AI-V30

[illegible] Bing Chat [illegible] [illegible] (2023)

Creative Problem Tackling – Methodology and Material of the East German Creativity Training Centre

Jochen Wessner(✉)

Esslingen, Deutschland
jochen.wessner2@de.bosch.com

Abstract. Effective and inventive problem solving is an important task. In the former German Democratic Republic, a training scheme was created by engineers and psychologist to deal with problems in a creative way to find new and patentable solutions. The training material and additional literature is still available at university and public libraries. The research question, this paper tries to answer, is, does the training still provide a systematic approach for today's problem situations and does it offer up to date tools to come up with suitable solutions. Furthermore, a short comparison to TRIZ is given.

Keywords: Inventive Problem Solving · Creativity · Methodology · TRIZ

1 Introduction

Problem analysis and solving is an important task and this is mostly still carried out by humans. Research showed that the level of creativity and creative problem solving could be improved by training [1, 2 p. 3]. Therefore, to foster the ability and effectiveness to solve problems an educational scheme was created in the former German Democratic Republic. The training was rather extensive and accompanied by a series of supporting booklets to provide material for self-study and looking up. The participants were mainly engineers and people with a technical background. The structure of the training consisted of moderated work in larger groups with skilled inventors and psychologist accompanied by self-training. Every participant of the training was considered to prepare a real problem that needed solving. The final solution should be new and inventive with the aim to file a patent application by the trainee. Parts of this training material [2–19] and additional literature [20, 21] is still present in university or public libraries mainly in the former German Democratic Republic.

The research question, this paper tries to answer, is, does the training still provide a systematic problem approach to current problem situations and does it offer up to date tools to come up with inventive solutions. The self-study training material consists of a series of booklets which covers a great variety of topics, e.g., a workflow to structure the process and tools for analysing and solving the problem.

O. Mayer (Hrsg.): TRIZ-Anwendertag 2024,
https://doi.org/10.1007/978-3-662-70698-5_3

The booklets 1, 2/1 and 2/2 deal with the background and motivation of the training, i.e., the meaning of creativity, the creative product, the creative person and its improvement.

Figure 1 shows parts of soft cover front pages of the training booklet series. The DIN A4 format booklets are typewriter written and contain several hand drawn diagrams and figures. Due to different authors of the training material the motivation to introduce this training was to achieve a systematic, efficient and highly creative problem-solving process leading to patentable solutions. The created intellectual property could also procure valuta to import needed raw materials.

Fig. 1. Soft cover front pages with the typical layout of the booklet series.

The authors perceive the problem tackling process as an information processing procedure (see Fig. 2). The inputs are tasks, known insights, methods and circumstances, the outputs solutions and new insights. It also states the interconnection between human and non-human, i.e., computer supported or initiated processes.

2 The Aim of the East German ctc-Trainings

Figure 3 shows the steps of a systematic problem-solving process. The process starts with the analysis of a given task to derive a precise definition of a task, i.e., the specific problem which is not necessarily the problem statement of the project owner. The specific problem is transformed into an abstract model of the problem, solved and from the abstract solution a specific solution is obtained. The creation of abstract problem models helps to overcome psychological inertia and to gains new perspectives.

The term creativity could be defined as the will to create something new. This personal attitude is supported by initiative and skill. Furthermore, it should be combi-

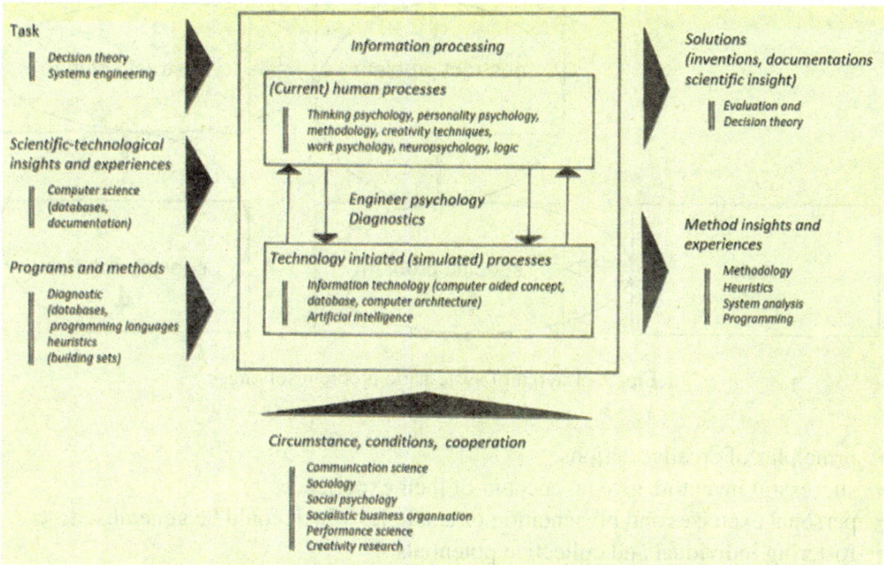

Fig. 2. Problem solving perceived as information processing [2, p. 6].

ned with the application and/or discovery of methods to give creativity a direction and come up with solutions in a systematic and efficient way.

Within the training ideally this produces an insight which leads to an internal behaviour during the process. The ctc-training aimed at improving the scientific-technical creativity because they found out that creativity could be trained and learned, i.e., acquired by training.

To support the process method specialists and senior development engineers are present helping with

- elements
- tools (material or immaterial)

in order to create value and exploit this value.

To support the psychological process psychologists are present during the training as well to deal with:

- personality (cognitive, emotional, motivational)
- thinking
- perception
- conduct, behaviour

2.1 The Training Motivation and Concept

The aim of the training was to raise the interest in creative action and to improve the quality of the solutions to achieve patentable inventions. Creative thinking and acting are learnable, therefore the course covers contents like:

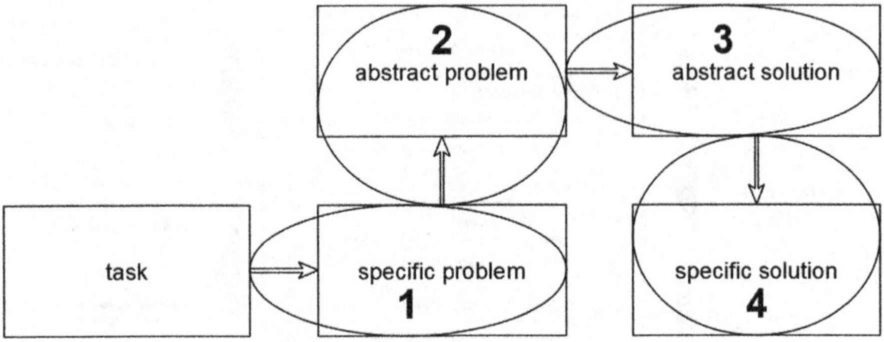

Fig. 3. Layout of systematic problem solving.

- principles of creative actions
- successful inventors give an account of their experiences
- personal exercises and presentation of examples which could be generalised
- fostering individual and collective potentials
- presenting possibilities to reduce obstacles which hinder the process

Therefore, the training addressed different aspect (see Fig. 4). It was installed to come up with creative and effective problem solvers who acquired an intrinsic motivation. Ideally, the solutions could be patented and generate value through licensing. The timeline of the training is shown in the third column form the left. After an opening session four intensive trainings consisting of self-study, consultation, seminars, three of them weeklong, followed and a final defence concluded the training. The timeline reflects the workflow of the participant's project. The booklets are mapped to this workflow (see Fig. 4, 1st column). The 4th and 5th column are references to training steps and the intended results from those steps. To support a systematic solution process references to methodologies are given in column 6. The trainers being experienced inventors, engineers and psychologist.

The combination of training and tackling of an individual problem up to the point of writing a patent application helps to transfer the knowledge, skill and behaviour to other problem areas.

2.2 The Creative Product and the Role of Need

The creative product could either be material (piece of art, a (technical) part, ...) or immaterial (information, methods, different problem models to view the problem from many different perspectives), e.g., a new, clever, efficient and patentable solution (product and process) which could be introduced without much delay and provide high benefit. The motivation for a creative product is need. Needs could be divided into employer, employee or social needs.

Employer or company needs are:

- fast and efficient solution to a problem
- gain profit and/or advantages over possible competitors
- increased number of patent applications (growth of intellectual property)

Employee or personal needs of a creative person could be [4, p. 10]:

- work in a creative way
- information, new insights and inspiration
- personal development and career
- security, stability, certainly, (reliability)
- belonging
- respect and appreciation
- independence (working freely, choose tasks)

Social needs include both aspects above and address for example accommodation, provision, public transport, health system and legislation.

2.3 The Creative Process and Supporting Materials

The result of the training was supposed to be an inventive solution to a real current problem which could be filed as a patent application. To achieve this the trainees should bring along a demanding task of their current work to the training and set up an in interdisciplinary team of 8–10 participants for collective and individual work. The teamwork allowed for analysing team building processes and the psychological issues of team work as well. The focus consisted of problem solving on the level of principle solutions and derive at a contradiction to be solved. The process should change between systematic and intuitive approaches, was guided by one methodologist and one psychologist. As stated, the training was supported by 17 continuously updated training booklets covering the complete process. Ideally the trainee helped to spread the knowledge in own trainings and to apply special findings which were gained during the training.

The process contained the following rough steps [2, p. 14; 20, p. 11]:

- Moving from a task to a problem statement and after an analysis step to a precise definition of the task (the abstract problem)
  - What is needed?
  - What is the subject of the problem (product, process, method, ...)?
  - What orientation should the solution provide?
    new and inventive
    improvement with an insightful compromise
    sufficient, acceptable or adequate solution
  - Analysis of the complexity of the solution and the amount of restrictions
- Solution finding
- Solution evaluation
- Tests and evaluation
- Postprocessing and final report

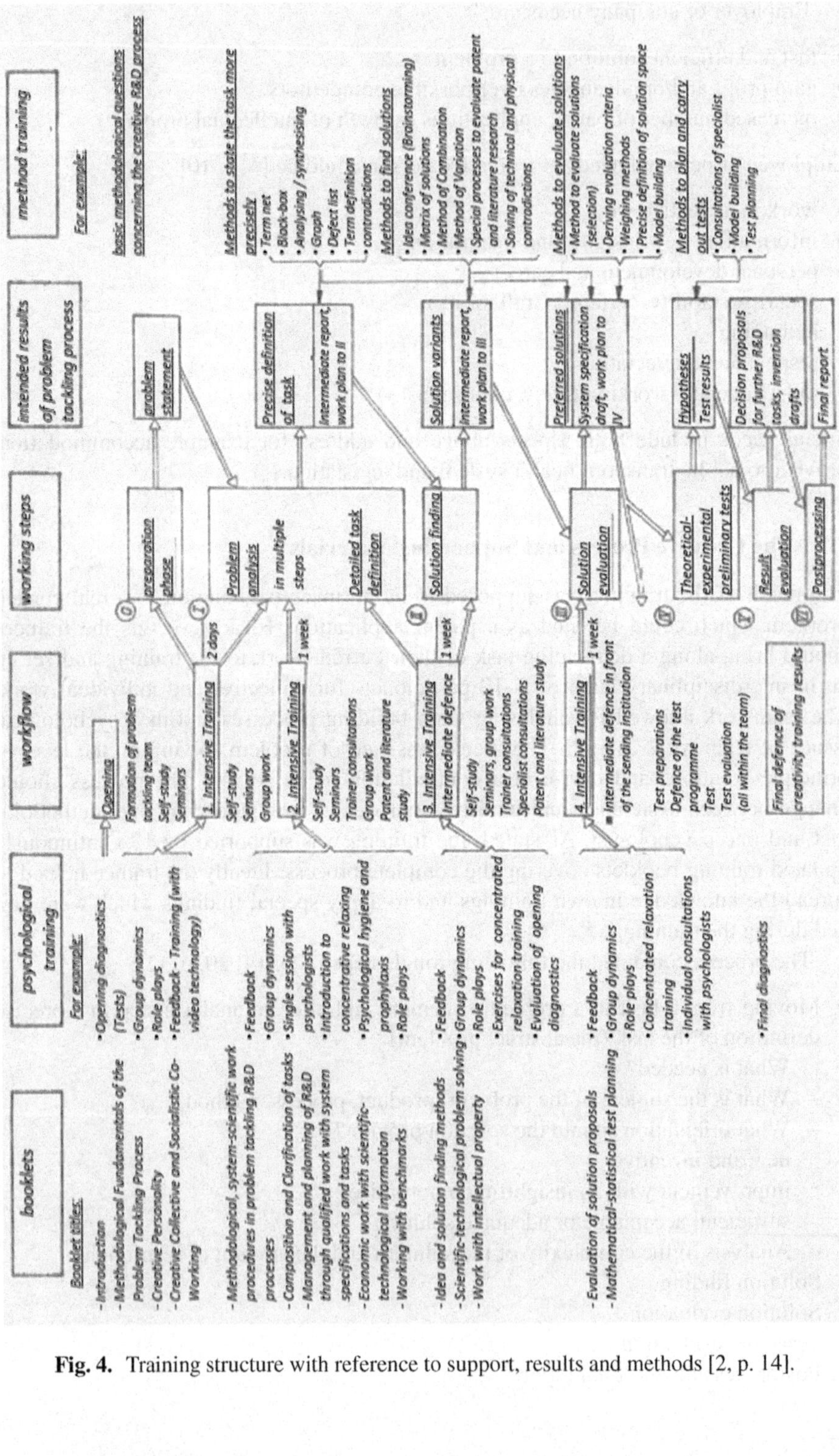

Fig. 4. Training structure with reference to support, results and methods [2, p. 14].

It is interpreted as a data processing procedure [2, p. 6]. Present, obtained and created data is turned into information and used in mental process steps to come up with a solution idea. This process requires an improvement or development of communication skills as well. Therefore, reading (training material, patents, journals, scientific literature), writing (report, requirement list, product documentation, own patent application-invention reports) and oral communication skills were addressed as well. Figure 5 shows support material presented in the booklets to analyse a text and in [22] Müller, one of the trainings methodologists, presented a guideline to write scientific texts in a structured way.

2.4 The Creative Person

The subject creating a creative product is the creative person. This person should combine different characteristics [3/1, pp. 10–13]:

- intellectual prerequisites
  - knowledge (acquired in different institutions (school, university), training)
  - skill (flexible way to choose what to apply)
  - capability and experience (being able to apply knowledge to a specific situation)
- character
  - motivation (intrinsic stimulus)
  - attitude (the will to apply knowledge to a specific situation and/or solve a problem).

These character qualities are demonstrated by the following properties:

- interest in finding, precepting and searching for contradictions in nature, technology and society
- reasonable doubt in obvious appearing principles and the ability to formulate constructive criticism and precise statements

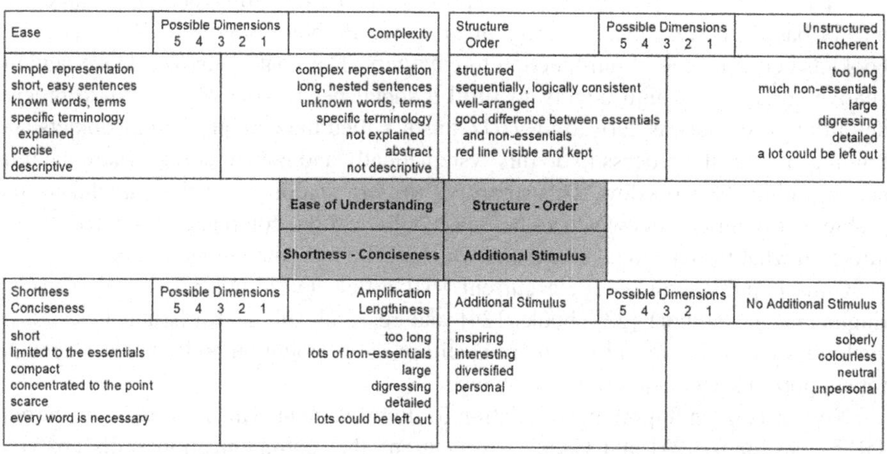

Fig. 5. Four dimensions of a text (arranged by the author) [3/2, pp. 11–15].

- ability to think and formulate alternatives, possessing phantasy, fluent thinking
- high identification with scientific work in general and especially when dealing with difficult problems
- dealing with divergent ideas and opinions (in written and oral form)
- willingness to take reasonable risks
- courage to have an isolated opinion and to see things from different (new, other) perspectives
- interest in thought experiments to systematically extend (enhance) the personal knowledge and experience.

The training was set up to train and improve these properties and behaviours and have an effect on others (colleagues, teams, superior, …).

Nevertheless, the training material stated that these personal characteristics could also appear as socially unadopted, self-centred, unbalanced and uncontrolled or not disciplined because thinking in a different way or having a different perspective is not always pleasing for the environment. (a negative side or perception _be negative or appear negative-possible contradiction). Instead of fostering teamwork or collective this would have the opposite effect and possibly create conflicts.

It was also thought that incentives (material and immaterial) are necessary to stimulate the inventive activity and for example the amount of money which could be gained from a so-called economic patent could be quite substantial compared to normal incomes in the former German Democratic Republic. In addition to money inventor medals were awarded for exceptional inventors.

3 Results

The third edition of the training material dates back to 1986 and is authored by people living in the former German Democratic Republic. Hence, it contains also elements of the socialistic doctrine, e.g., the work in combinates.

Apart from those elements, the approach of tasks is still very up to date and mostly based on Altshuller's Theory of Inventive Problem Solving (TRIZ). The first booklets act as a general introduction to creativity. The goal being to achieve new inventive solutions of high quality to fulfil a social need. After an analysis a precise definition of the task is derived, in TRIZ this is called the core problem. Booklets and training support the process to do this systematically and help to define where the trainee is within the workflow. This improves the orientation and helps structuring the problem-solving process which sometimes could be quite complicated. It also offers a direction what to steps to take next and which of the presented tools to use.

Current publications, e.g., a current VDI article (VDI: Association of German Engineers, the author) [23], books [24] and research of the Karlsruhe Institute of Technology [25, 26] reveal that there is still research going on and a present demand for support in systematic problem solving.

Need acts as an important stimulation for improvement. This is stressed by current TRIZ consultants [27] and Gadd also mentions that communication skills are vital within project work [28]. The training covers this by mandatory reading of patent publications and scientific texts, writing project documentation and a final report plus oral defences of found results. The foundation of the above is the result of treating the problem-solving process as processing data in a mentally way.

The duration of the training, at least three and a half weeks, is quite long and might result in implementation problems, i.e., the trainee is not present at work. Taking into account the training contents and the fact that current TRIZ trainings up to Level 3 take up the same amount of time this duration seems to be required to achieve a solid training effect and ensure the application of the acquired skill and knowledge to practical problem situations. The longer the methodology is trained the more habitual it becomes in situations when methodological approaches are useful or required.

The different trainers, being experienced inventors, engineers, scientists and psychologists who developed the creativity training based on their research, provide different perspectives and help to gain a broader view and help to overcome deadlocks which could arise due to outside pressure or difficult surroundings. The training material set rules and the order for their application is coordinated with the workflow of a project. Several tools, e.g., questionnaires and schematics resemble TRIZ tools (process analysis, defect list, black-box diagram) or are carried over directly (definition of a contradiction and solving with 35 inventive principles, ARIZ, Substance-Field Modelling). Short cuts are hinted for smaller projects or tasks as well because the methodology should not be an end in itself.

The trained skills and and the training booklets seem not to be limited to a specific problem situation. They are very generic, and the topic is thought to be very up to date. The application does not ensure 100% reliability to find a solution, but it is a considerable improvement to guessing and unsystematic actions. The increase of the trainees' level of creativity could be measure as well if needed. The number of created problem models or different perspectives considered offer a good correlation to an increased level of creativity or problem-solving skill. The more problem models the more solutions options are created. This could result in an increased number of patent applications as well.

In [29] a first computational support is hinted. The current computer processing power and easy accessibility of public databases, e.g., patent publications and internet articles, effect databases increased this support substantially.

The goal of the training was that each participant filed a patent publication of a current or pressing problem. This makes carrying over the training set up to current trainings difficult. Competition and confidentiality could restrict the application for cross company trainings and extensive exchange of thoughts or solution ideas.

At the end of each booklet a reference to additional literature could be found, e.g., to TRIZ, NOIS (Nature oriented innovation strategy) [30, 31], Systematic Heuristics or a similar training programme [32, 33].

4 Summary and Conclusions

Effective and inventive problem solving is an important task. The training programme created in the former German Democratic Republic is therefore far from outdated. The different perspectives offered for the trainee by a wide range of trainers and their diverse background should enable the participant to deal with problems in a systematic and creative way to find new, inventive and hence patentable solutions. The training material and additional literature is found to be very useful within the problem-solving process and available at university and public libraries; although it needs some effort to procure it. The presented tools help to gain new ideas and to give some orientation within a sometimes-unclear project situation. The short comparison to TRIZ shows that the creativity training and TRIZ are partly identical or have very much in common, e.g., making a problem more abstract and find an abstract solution to the abstract problem model which is then made specific again.

References

1. Altshuller, G.S.: Creativity as an Exact Science. Gordon and Breach, New York (1995). ISBN 0-677-21230-5
2. Heyse, V., Bausdorf, J.: Lehrbrief 1 – Einführung in die Lehrbriefreihe. In: Lehrbriefreihe Grundlagen des wissenschaftlich-technischen Schöpfertums in Forschungs- und Entwicklungsprozessen. Bauakademie der DDR, 3. Aufl. Jena, Berlin (1986)
3. Metzler, H.: Lehrbrief 2 Teil 1 – Methodologische Grundlagen des schöpferischen Problemlosungsprozesses. In: [2]
4. Thiel, R.: Lehrbrief 2 Teil 2 – Methodologische Grundlagen des schöpferischen Problemlosungsprozesses, Was ist „Methode?". In: [2]
5. Heyse, V.: Lehrbrief 3 – Kreative Persönlichkeit im Forschungs- und Entwicklungsprozess. In: [2]
6. Heyse, V., Buck, M., Pastillé, G.: Lehrbrief 3 Zusatzmaterial – Zur Theorie der Höchstbelastung. In: [2]
7. Ladensack, K., Benes, J.: Lehrbrief 4 – Schöpferisches Kollektiv und sozialistische Gemeinschaftsarbeit. In: [2]
8. Koch, P.: Lehrbrief 5 – Methodisch-systemwissenschaftliche Arbeitsweise bei der Problemlösung in Forschungs- und Entwicklungsprozessen. In: [2]
9. Koch, P.: Lehrbrief 6 – Ausarbeiten und Präzisieren von Aufgabenstellungen. In: [2]
10. Gramsch, H.-U., Preisler, W.: Lehrbrief 7 Teil 1 – Rationelle Arbeit mit wissenschaftlich-technischer Information, Grundlagen. In: [2]
11. Gramsch, H.-U., Grün, F., Förster, F.-U., Spindler, R., Vespermann, G.: Lehrbrief 7 Teil 2 – Arbeit mit spezifischen Informationsquellen Bauwesen. In: [2]
12. Sabisch, H., Franke, S., Zanger, C.: Lehrbrief 8 – Arbeit mit Weltstandsvergleichen. In: [2]
13. Sohn, M.: Lehrbrief 9 – Leitung und Planung von Forschung und Entwicklung durch qualifizierte Arbeit mit Pflichtenheften. In: [2]
14. Mehlhorn, H.-G., Gutzer, H.: Lehrbrief 10 – Ideen- und Lösungsfindungsmethoden. In: [2]
15. Busch, K.: Lehrbrief 11 Teil 1 – Lösungsermittlung wissenschaftlich-technischer Problemstellungen. In: [2]
16. Preisler, W.: Lehrbrief 12 – Bewerten von Lösungsvorschlägen. In: [2]
17. Busch, K.: Lehrbrief 13 – Mathematisch-statistische Versuchsplanung. In: [2]
18. Zumpe, V.: Lehrbrief 14 – Schutzrechtsarbeit. In: [2]
19. Springer, R.: Lehrbrief 16 – Marktarbeit für die Entwicklung neuer Erzeugnisse und Verfahren und deren Export in das NSW. In: [2]

20. Heyse, V., Bausdorf, J., Busch, K., Koch, P., Rindfleisch, J.: Vorbereitungsmaterial für internationale Spezialkurse „Kreative Bearbeitung technisch-naturwissenschaftlicher Probleme in interdisziplinären Gruppen". Bauakademie der DDR. Berlin (1987)
21. Rüdrich, G., Grünberg, H.-U.: Nutzung von naturgesetzmäßigen Effekten und Wirkprinzipien zur kreativen Bearbeitung technisch-naturwissenschaftlicher Probleme. Bauakademie der DDR. Berlin (1988)
22. Müller, J.: Arbeitsmethoden der Technikwissenschaften. Springer, Berlin (1990). ISBN 978-3-642-93443-8
23. Koppenhagen, F., Wecht, C. H.: Warum das nächste iPhone wieder nicht aus Deutschland kommt: Der Zusammenhang zwischen Entwicklungsmethodik und radikalen Produktinnovationen. VDI Fachmedien Konstruktion (2023) Nr. 6. https://doi.org/10.37544/0720-5953-2023-06-68
24. Conn, C., McLean, R.: Bulletproof Problem Solving. Wiley, New Jersey (2018). ISBN 9781119553021
25. Nelius, T., et. al.: Tatort Technik (Teil 1) – Herausforderungen bei der Problemanalyse in der Konstruktion. VDI Fachmedien Konstruktion (2021) Nr. 3. ISSN print: 0720-5953
26. Nelius, T., et. al.: Tatort Technik (Teil 2) – Unterstützung bei der Problemanalyse in der Konstruktion mit der Design-ACH-Methode. VDI Fachmedien Konstruktion (2021) Nr. 4. ISSN print: 0720-5953
27. Fey, V., Rivin, E.: Innovation on Demand, 6th edn. Cambridge University Press (2012). ISBN 978-0-521-82620-4
28. Haines-Gadd, L.: TRIZ for Dummies. Wiley, Padstow (2016). ISBN 978-1-119-10747-7
29. Herrig, D.: Programmpaket HEUREKA – Hilfsmittel zum Erfinden unter Rechner- und Karteinutzung. Bauakademie der DDR (1988)
30. Hill, B.: Innovationsquelle Natur. Shaker, Aachen (1997). ISBN 3-8265-2887-5
31. Hill, B.: Erfinden mit der Natur. Shaker, Aachen (1998). ISBN 3-8265-4139-1
32. Zadek, G., Herrlich, M.: KDT-Erfinderschule Lehrmaterial 1. Teil. KDT-Eigenverlag, Berlin (1982)
33. Zadek, G., Herrlich, M.: KDT-Erfinderschule Lehrmaterial 2. Teil. KDT-Eigenverlag, Berlin (1982)

Creative Problem Tackling – Problem Analysis (East German Creativity Training Centre Methodology)

Jochen Wessner(✉)

Esslingen, Deutschland
jochen.wessner2@de.bosch.com

Abstract. The paper tries to answer the question whether this part of the training material supports finding a precise problem definition. The selected question schemes and workflow examples suggest that the training is still very useful to deal with problems in a creative way. Especially moving from the desired result to inputs, circumstances and side effects together with moving from the system border to the inside of the system helps to structure the first approach to a problem situation. This facilitates identifying defects, i.e., information shortcomings and getting over various bias. Furthermore, the diagrams support the planning of a solving process by showing various influencing aspects which should be taken into account and also assist TRIZ analysis like resource and process analysis.

Keywords: Problem-solving · Problem analysis · Creativity · TRIZ

1 Introduction

In [1–20] a creativity training is described. This training was developed in the former German Democratic Republic to improve the level of creative and effective problem solving. The goal was to come up with inventive and patentable solutions to problems with social need for solving, e.g., satisfy demands, sell licenses and raise valuta. Some booklets [5–8, 11] focus on the aspect of "What has to be done to derive an analysed precise problem definition from a given task?" Solving the right problem is a key factor in the development process. Figure 1 shows a graph in which the caused costs of a product (in percent of the total cost) is plotted over the percentage of the defined costs with respect to the different product life stages. The graph states that roughly 80–90% of the final cost of a product is defined during the first two stages, i.e., during the development (design) and the chosen technology. This statement is rather old but newer publications [21, 22] show similar diagrams. Therefore, solving the right problem is essential because later changes consume much more effort. The question this paper tries to answer is, does this training material helps to find a precise problem definition.

O. Mayer (Hrsg.): TRIZ-Anwendertag 2024,
https://doi.org/10.1007/978-3-662-70698-5_4

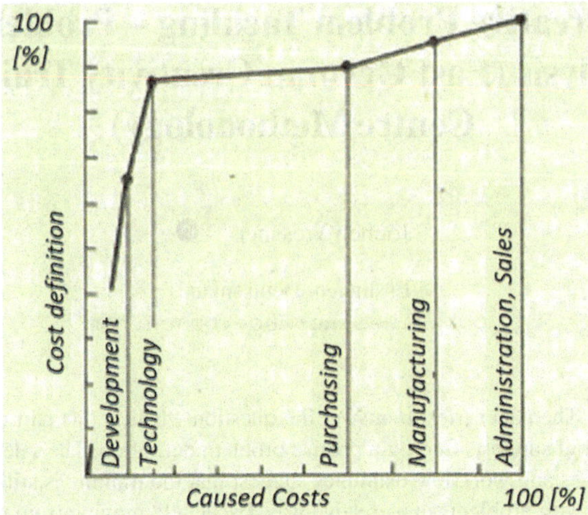

Fig. 1. Defined costs over caused costs for different product lifecycle stages [5, p. 1]

2 Moving from a Task to a Key Problem

Tasks could arise from various issues, typical ones are for example, new product variants, product improvement or completely new products. These tasks or improvements aims are not always clearly defined, sometimes even misleading. Zobel calls these task poisonous tasks [23, p. 44] due to the fact that posing and accepting assumptions without further questioning results in creating a possible perception bias which might cause psychological inertia and unnecessary or even dangerous requirements and solution constraints.

For getting a clearer view on what has to be done it must be defined on what level the change or improvement should take place. Moving upwards on the left-hand side, Fig. 3 shows different levels of development tasks and the interrelation between the aspect of function, dealing with what must be achieved and the structure, dealing with how to achieve it. The newer the product or greater the change the higher the effort and the more important to cover all stages of the product lifecycle (e.g., development, manufacturing, sales, usage and rejection) since different materials or new production processes have a strong influence on cost. The term "product" is not restricted to physical goods but could be, with some abstraction and generalisation, extended to services, business models or software as well. After arriving at the desired level, a transition is stated. This movement from the left column to the right one is accompanied with a transition from purpose, system border and function to the needed means for fulfilling the purpose, namely with the system interior and with system structure. of change the movement. Furthermore, the patentability of inventive solutions is hinted (Fig. 2).

Figures 3 and 4 show the system as a black box representation together with questions which should be answered to precisely define the task that should be carried out. The first of the two diagrams displays a general layout and the numbers state an order in which order the questions should be answered. Notably, the first question deals with the desired result which is to be achieved. This strongly resembles a TRIZ workflow, i.e., starting with the definition of the ideal final result [24, 25], and other workflows in current problem-solving literature [26].

Question 2, "What is present?", supports identifying present or derivable resources and question 3 identifying constraints which have to be met. Additional questions and examples of possible answer aspects is illustrated in Fig. 4. All of the above aims at defining a precise description of the problem. Possible biases like perception bias, priming or anchoring [27, 28] could be overcome.

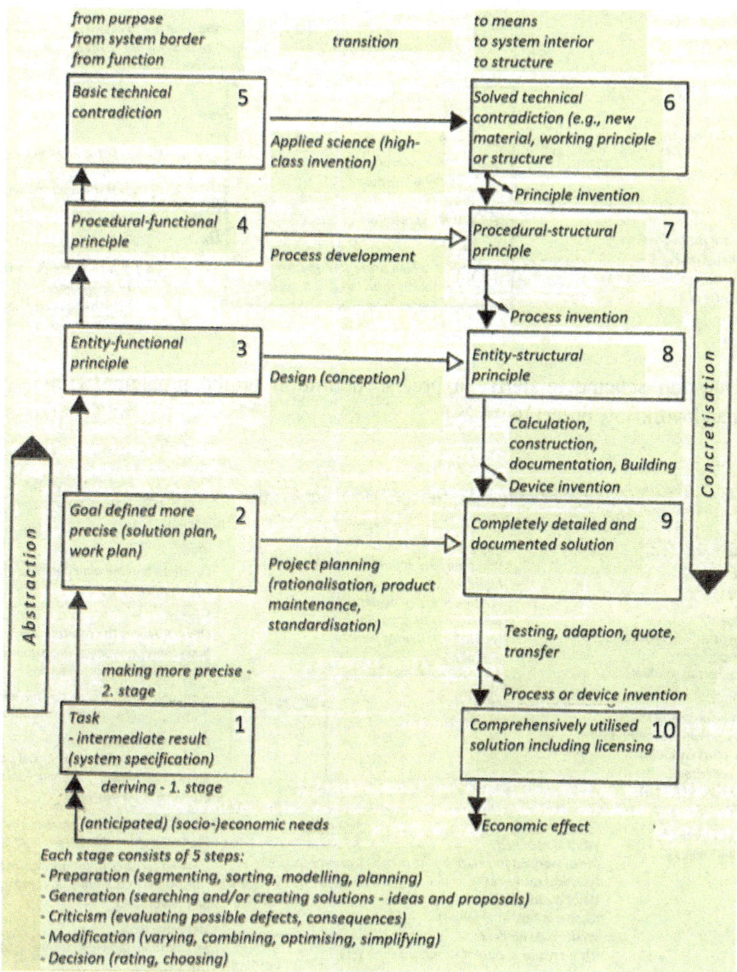

Fig. 2. Deriving the level of abstraction of the problem and tasks to be done [11, p. 10]

The training treats the problem-solving process as a procedure of data processing. Therefore, if needed data or information is not present it must be created.

Anything which is not present, misleading or unclear should be indexed and listed in a defect list. In [6, p. 58–64; 29] an example of such a defect list is given, thereby showing the underlying complexity of a simple looking task (volume increase of a machined metal part – milling process).

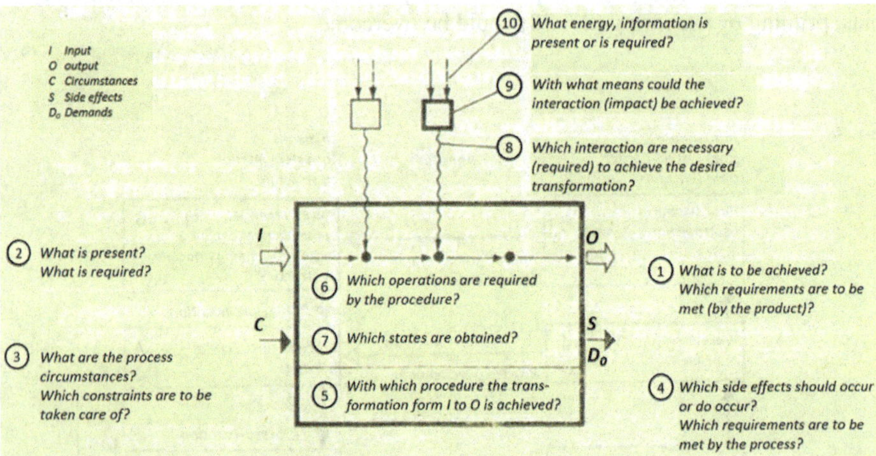

Fig. 3. Question scheme to derive a precise definition of the problem (with intermediate operands and workflow order) [6, p. 29]

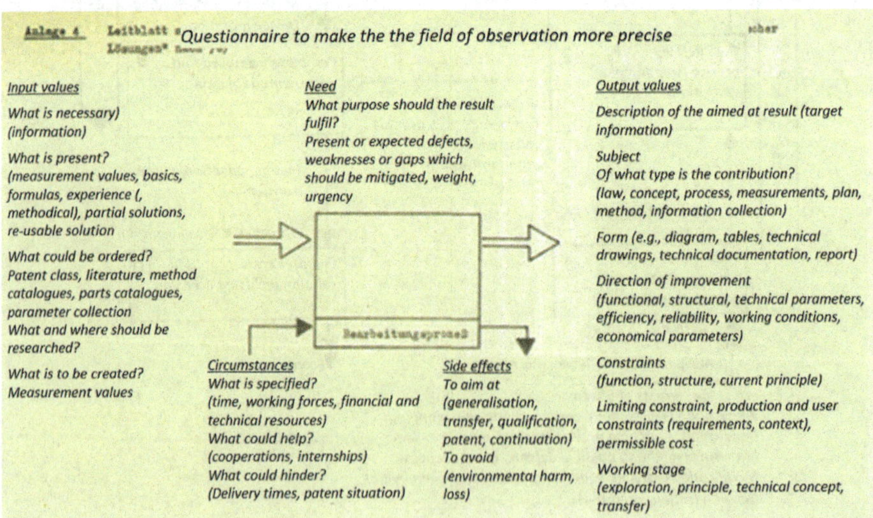

Fig. 4. Question scheme to derive a more precise definition of the problem (with focus to the environment) [6, p. 41]

Figure 5 displays the problem-solving workflow moving from the working principle to the functional principle during a concretisation phase. It contains 8 general steps which the authors of the training material believe to be invariant and hence could be used for other problem-solving processes as well.

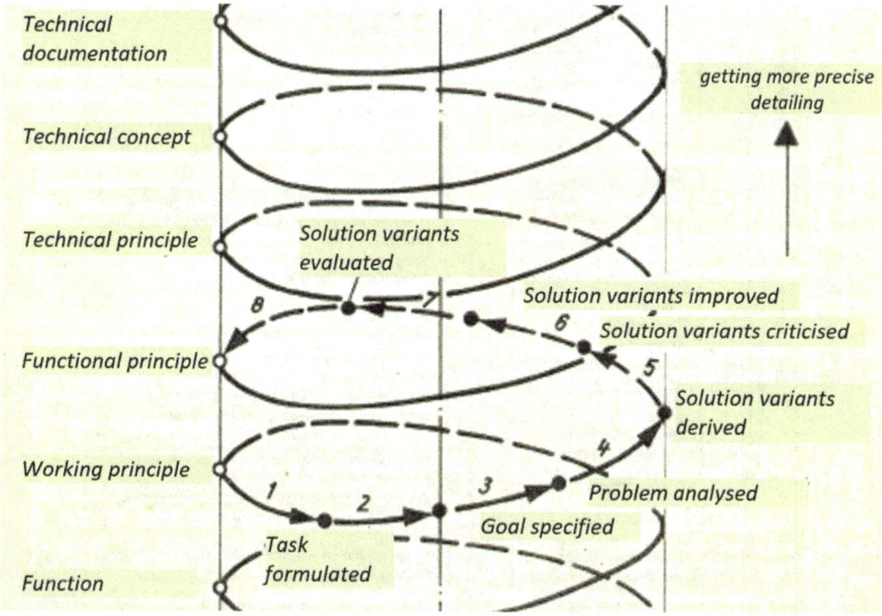

Fig. 5. General stages of a problem tackling process in direction of concretising/detailing [5, p. 47]

The next figure, Fig. 6 illustrates this workflow in more detail. The actual solving, step number 4 is only one of the 8 steps. A strong focus lies on the problem analysis prior to solving and evaluating or improving solution variants. The arrows provide a direction and indicate possible iterations or short cuts. On the right-hand side of the figure characteristic intermediate results are given. These results, representing important information, should be documented.

Iterations, omissions, loops and other process variations strongly influence the planning of a problem tackling process. Figure 7 provides an overview of the great complexity and highlights aspects of this planning process. Starting from a sequential workflow in the centre of the diagram influences are shown in an outer loop like structure. These influences are:

- classification: phases, deliverables, process structure
- specification: complicatedness and complexity of the object
- modification: omission, anticipation, repetition, loop and parallelism
- detailing: sequence of operation, process structure and serial or parallel sub-steps.

The screw like movement shows that these steps appear again at the beginning of the following stage, e.g., moving from functional to technical principle.

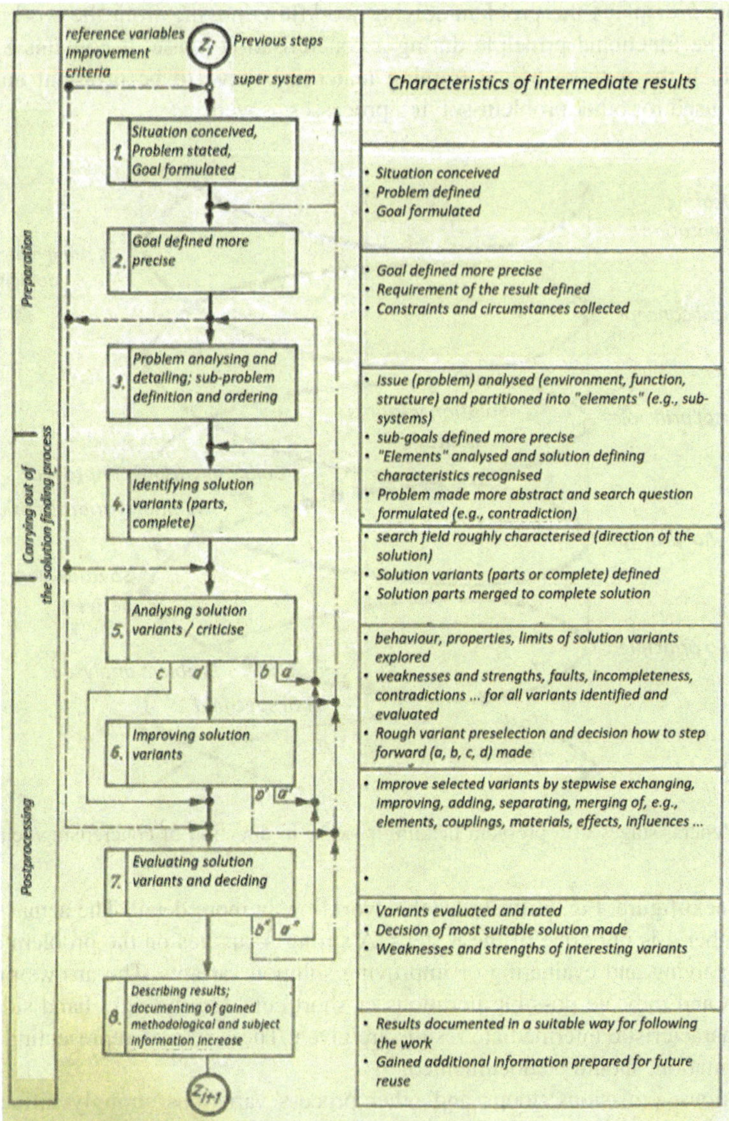

Fig. 6. Workflow of a problem tackling process [5, p. 48]

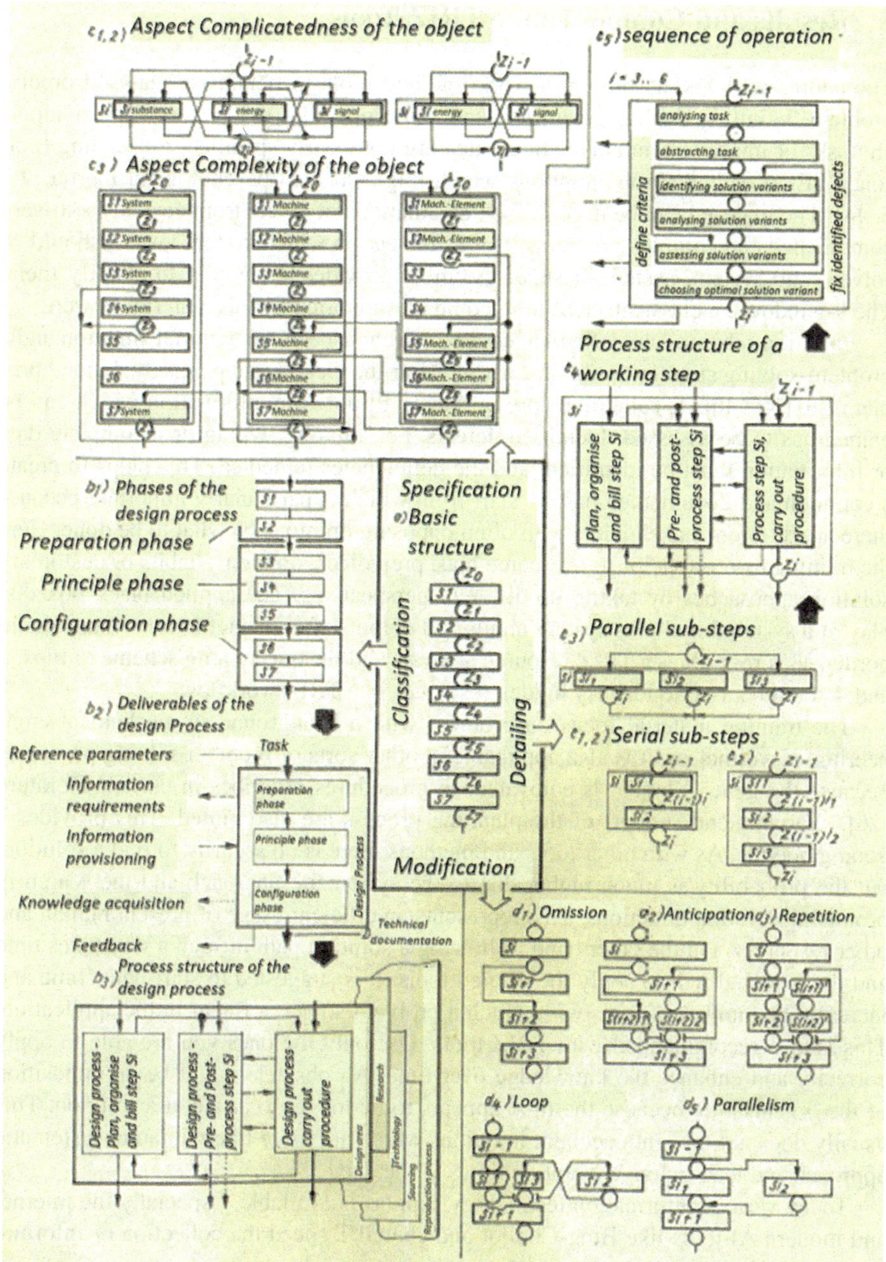

Fig. 7. General aspects of a problem tackling process – its operations Z_i and its planning steps S_i [5, p. 31]

3 Results and Comparison to TRIZ Tools

The addressed booklets present a workflow and tools to obtain an analysed precise problem definition from a given task and some planning guidelines. Typical inputs are tasks or improvement aims which often are not clearly specified, misleading (poisoned tasks) or contain assumptions which might lead to psychological inertia [23, p. 44]. The material can be used to look at situations of levels from easy tasks to very complicated and complex problems. If it is unclear on which system level it should be solved (sub system, system or supersystem) it provides some help to classify them. The 9-windows, a classical TRIZ tool, could be supported by this material as well.

Question schemes are defined to cover different aspects of an initial situation and a problem-solving strategy could be derived. After the clearly and precisely defined problem (in TRIZ this is called the core problem) all possible information which can be gained has to be assessed. Identified defects, i.e., missing, wrong or incomplete data or information must be identified, and the deficiencies remedied. This helps to create a common and documented base to start from which is particularly important because there are different stakeholders with often opposing opinions on what to be done. Here the training material helps to overcome bias, prejudices, opinions, habits or customary solution approaches by taking up different perspectives. The applied black box display of a system highlights system inputs and outputs after the definition of the system border. This resembles a TRIZ resource analysis and the questioning scheme in Figs. 3 and 4 could be used to identify hidden resources in a TRIZ workflow.

The training material focus on problem with a background in mechanical engineering. Nevertheless, it is also applicable in other sorts of problem-solving processes because the general layout is equivalent to procedures describes in current literature [26]. Variants and support of the planning process are also hinted. This provides a good guideline. As with other tools and methods there is no security to find a solution, but the probability is much higher due to the systematic approach and the search in promising solution directions. This represents an efficient usage of present human and other resources. On the other hand, following a solution path in such a way takes time and the personal will to apply the offered tools. It is suggested to start with some and increase the number of tools with which a problem solver is fluent in the application. This is also recommended with TRIZ tools. Use only the ones you are able to apply correctly and enhance the knowledge over time. An obstacle might be an opposition of the organisation because the time appears to be longer to arrive at a solution. This usually does not take into account iterations which are more likely in an unsystematic approach and lead to longer solution times in the end.

Today various information technology support is available. Especially the internet and modern AI tools like Bing-Copilot and chatGPT speed the collection of information up or help organising present information in a suitable way.

4 Summary and Conclusion

The paper tries to answer the question whether this part of the training material supports finding a precise problem definition. The selected question schemes and workflow examples suggest that the training is still very useful to deal with problems in a creative way. Especially moving from the desired result to inputs, circumstances and side effects together with moving from the system border to the inside of the system help to structure the first approach to a problem situation. This helps identifying defects and overcome various bias, e.g., perception bias and question assumptions stated by the problem owner and involved stakeholders. Furthermore, the diagrams support the planning of the solving process by showing various influencing aspects which should be taken into account.

The abstract character of the material sometimes makes it difficult to use but the treatment as an information processing procedure facilitates the transfer from problems with a mechanical engineering background to a general problem-solving process. Unavailable information, which is called a defect, should be procured. Other booklets not addressed in the paper support this process [9–11].

In general, it can be said that the steps to arrive at a clearly specified and analysed problem formulation bear a great resemblance to TRIZ. Particularly the first steps of ARIZ 68 and 71, resource analysis and process analysis are very similar to the steps shown in the training material.

References

1. Heyse, V., Bausdorf, J.: Lehrbrief 1 – Einführung in die Lehrbriefreihe. In: Lehrbriefreihe Grundlagen des wissenschaftlich-technischen Schöpfertums in Forschungs- und Entwicklungsprozessen. Bauakademie der DDR, 3. Aufl. Jena, Berlin (1986)
2. Metzler, H.: Lehrbrief 2 Teil 1 – Methodologische Grundlagen des schöpferischen Problemlosungsprozesses. In: [2]
3. Thiel, R.: Lehrbrief 2 Teil 2 – Methodologische Grundlagen des schöpferischen Problemlosungsprozesses, Was ist „Methode?“. In: [1]
4. Heyse, V.: Lehrbrief 3 – Kreative Persönlichkeit im Forschungs- und Entwicklungsprozess. In: [1]
5. Heyse, V., Buck, M., Pastillé, G.: Lehrbrief 3 Zusatzmaterial – Zur Theorie der Höchstbelastung. In: [1]
6. Ladensack, K., Benes, J.: Lehrbrief 4 – Schöpferisches Kollektiv und sozialistische Gemeinschaftsarbeit. In: [1]
7. Koch, P.: Lehrbrief 5 – Methodisch-systemwissenschaftliche Arbeitsweise bei der Problemlösung in Forschungs- und Entwicklungsprozessen. In: [1]
8. Koch, P.: Lehrbrief 6 – Ausarbeiten und Präzisieren von Aufgabenstellungen. In: [1]
9. Gramsch, H.-U., Preisler, W.: Lehrbrief 7 Teil 1 – Rationelle Arbeit mit wissenschaftlich-technischer Information, Grundlagen. In: [1]
10. Gramsch, H.-U., Grün, F., Förster, F.-U., Spindler, R., Vespermann, G.: Lehrbrief 7 Teil 2 – Arbeit mit spezifischen Informationsquellen Bauwesen. In: [1]
11. Sabisch, H., Franke, S., Zanger, C.: Lehrbrief 8 – Arbeit mit Weltstandsvergleichen. In: [1]
12. Sohn, M.: Lehrbrief 9 – Leitung und Planung von Forschung und Entwicklung durch qualifizierte Arbeit mit Pflichtenheften. In: [1]
13. Mehlhorn, H.-G., Gutzer, H.: Lehrbrief 10 – Ideen- und Lösungsfindungsmethoden. In: [1]
14. Busch, K.: Lehrbrief 11 Teil 1 – Lösungsermittlung wissenschaftlich-technischer Problemstellungen. In: [1]
15. Preisler, W.: Lehrbrief 12 – Bewerten von Lösungsvorschlägen. In: [1]
16. Busch, K.: Lehrbrief 13 – Mathematisch-statistische Versuchsplanung. In: [1]

17. Zumpe, V.: Lehrbrief 14 – Schutzrechtsarbeit. In: [1]
18. Springer, R.: Lehrbrief 16 – Marktarbeit für die Entwicklung neuer Erzeugnisse und Verfahren und deren Export in das NSW. In: [1]
19. Heyse, V., Bausdorf, J., Busch, K., Koch, P., Rindfleisch, J.: Vorbereitungsmaterial für internationale Spezialkurse „Kreative Bearbeitung technisch-naturwissenschaftlicher Probleme in interdisziplinären Gruppen“. Bauakademie der DDR. Berlin (1987)
20. Rüdrich, G., Grünberg, H.-U.: Nutzung von naturgesetzmäßigen Effekten und Wirkprinzipien zur kreativen Bearbeitung technisch-naturwissenschaftlicher Probleme. Bauakademie der DDR. Berlin (1988)
21. Ehrlenspiel, K, et. al.: Kostengünstig Entwickeln und Konstruieren, 8. Aufl., S. 13. Springer Vieweg (2020). ISBN 978-3-662-62591-0
22. Bender, B., Gericke, K.: Pahl/Beitz Konstruktionslehre, 9. Aufl., S. 177. Springer (2021). ISBN 978-3-662-57303-7
23. Zobel, D.: Erfinderfibel. VEB Deutscher Verlag der Wissenschaften, 2. Aufl. Berlin (1987). ISBN 3-326-00131-2
24. Gadd, K.: TRIZ for Engineers. Wiley, Chichester (2011). ISBN 978-0-470-74188-7
25. Koltze, K., Souchkov, V.: Systematische Innovation, 2. Aufl. Hanser, München (2017). ISBN 978-3-446-45127-8
26. Conn, C., McLean, R.: Bulletproof Problem Solving. Wiley, Hoboken (2018). ISBN 9781119553021
27. Nelius, T., et. al.: Tatort Technik (Teil 1) – Herausforderungen bei der Problemanalyse in der Konstruktion. VDI Fachmedien Konstruktion (2021) Nr. 3. ISSN print: 0720-5953
28. Nelius, T., et. al.: Tatort Technik (Teil 2) – Unterstützung bei der Problemanalyse in der Konstruktion mit der Design-ACH-Methode. VDI Fachmedien Konstruktion (2021) Nr. 4. ISSN print: 0720-5953
29. Stanke, K.: Handlungsorientierte Kreativitätstechniken. trafo, Berlin (2011)

Application of TRIZ in a Hospital

Thomas Meindl(✉)

Fürth, Deutschland
t.meindl@ibom.de

Abstract. This paper deals with the topic of cost optimization in hospitals, particularly in the context of the economic pressure they are under. This pressure results, among other things, from the introduction of the DRG system, which has led to a focus on lucrative and simple cases. In order to save costs, hospitals are implementing various measures, including external measures such as merging hospitals or outsourcing services, as well as internal measures such as process optimization. One example of a process that can be optimized is patient transport planning within a hospital. Here, the sample process is analyzed and considered with the help of logistics approaches. Optimization opportunities can be identified by applying principles from logistics, for example from Deutsche Bahn. Another process analyzed is regular inpatient admission. This process is characterized by many influencing variables and disruptive factors. By applying innovation principles from TRIZ, ideas can be generated and prioritized in order to improve the process.

Keyword: TRIZ for Healthcare

1 Introduction [1–4]

Many hospitals, and in particular non-commercial hospitals, have been operating at a loss for several years and regularly require external financial support. This has been a major topic in the press, politics and society for years. Most experts see the main reason for this in the introduction of the DRG system (Diagnosis Related Groups). The DRG system assigns for every case a defined amount of money independent, if this case is simple (short hospital stay, no complications) or complicated (infections, poor recovery, long hospital stay.This means, that specialist clinics have been able to concentrate on lucrative and simple cases, while the more loss-making and complex cases have been referred to hospitals that have not been able to specialize enough or at all – such as municipal clinics or university hospitals or that must take the case.

As a result, hospitals, which have been under enormous economic pressure, especially since the above-mentioned reform, are aiming to save costs. However, it is difficult to specialize a hospital with a certain care mandate for the population. For this reason, such problems can be avoided, for example, by connecting smaller, rural hospitals to large maximum care providers, as complex cases are still treated by the same group and minor cases by smaller hospitals. This means that no capacity is tied up at

O. Mayer (Hrsg.): TRIZ-Anwendertag 2024,
https://doi.org/10.1007/978-3-662-70698-5_5

maximum care providers for complex cases. In addition to the cost savings, other advantages include, for example, reduced staffing levels by avoiding duplication.

In addition to these so-called "external" measures, there are also "internal" measures. These include outsourcing various services, such as the laundry, cleaning service or security service. As a result, such structures would not have to be set up in the first place or could be dissolved. These tend to be uneconomical on a smaller scale within a hospital. In addition, purchasing groups with other hospitals are possible, which has a scaling effect. This in turn leads to cost savings. Cost savings can also be achieved in terms of personnel through a lower nursing ratio or a lower number of doctors, as personnel costs in particular make up a large proportion of the costs incurred. Costs can also be reduced by purchasing cheaper products with the same function or by purchasing generic products.

There is also the option of process optimization. This is still rarely widespread or used, especially in hospitals. The author is a medical doctor and trained TRIZ L1. He applied TRIZ in his hospital environment. Various process optimization tools were explained in order to show in a next step, using several examples, how TRIZ can be used to optimize sample processes in hospitals and thus create a further tool in the cost optimization of hospitals.

2 Patient Transport Planning within a Hospital [5–7]

In smaller hospitals, such patient transports are often carried out by the ward or functional unit staff because the distances are usually relatively short. The situation is different in larger hospitals. The longer the distance to the destination, the longer the nurse is absent from the ward to continue caring for the other patients. For this reason, there is often an internal transport service with medically qualified staff, which, in addition to its main task – patient transport – often also takes on other important transport tasks within the hospital. First of all, the model process was defined:

1. First of all, the ward nursing staff must determine the need for transportation, for example due to a medically prescribed examination or therapy.
2. The date of the examination or therapy has been set and transportation is requested using the transport system.
3. A supervisor schedules the transport and assigns it to a member of the transport team. However, if other urgent transports are pending, this may result in delays.
4. After the patient is prepared, the transport is carried out. The nurse has made the patient ready for transport, for example by finishing infusions, washing the patient again if necessary and removing all unnecessary items, such as a bedpan.
5. The transport service hands the patient over to the receiving staff at the transport destination.

This process is handled with the help of the function-oriented search, whereby you can quickly find what you are looking for in logistics. For example, freight transport, which has to bring goods from A to B by defined deadlines. Or local public transport, Deutsche Bahn, airlines such as Lufthansa, ferries, etc. The business model of all these companies is based on transportation. In order to find the best model, several areas of logistics were considered, with Deutsche Bahn appearing to be the most promising. The next step is to find out what can be learned from them: What systems it uses, what algorithms, what software, what procedures to optimize transport. Of course, hospitals do not require the same level of complexity as rail transport. However, it can be assumed that one or two procedures could be copied and thus optimize the patient transport system in the hospital.

3 Regular Inpatient Admission [6, 7]

Another process that was analyzed is a regular inpatient admission. This process can be planned and is run through several times a day in order to admit patients who do not come to the hospital via the emergency department but with a medical referral. In detail, however, it is a process that has many influencing variables and disruptive factors due to the individualization of information for the patient.

The approach for such a complex process was to examine the 40 innovation principles for their benefits. These are shown in Table 1 in the appendix. In a further step, the ideas must be sorted and clustered in order to implement the best possible ones.

Table 1: Solution ideas generated with TRIZ

No.	Innovation principle	Solution ideas
1	Decomposition (segmentation)	The login process is broken down into individual segments that are processed asynchronously by different departments. The logon process is carried out in several stages
2	Separation	The registration process is carried out separately from the hospitalization process. Only when it has been successfully completed will the patient be admitted to hospital
3	Local quality	For the registration process, the patient is assigned a coach in advance who assists them in compiling the documents (also on site at home)
4	Asymmetry	No idea
5	Coupling	The family doctor compiles all the necessary documents with the patient and then hands them over to the hospital in full
6	Universality	A patient documentation system is being developed that makes the compilation of patient documentation superfluous because it is available and updated at any time
7	Integration	Closer cooperation between the hospital and the GP concerned

(Fortsetzung)

Table 1: (Fortsetzung)

No.	Innovation principle	Solution ideas
8	Counterweight	No idea
9	Previous counteraction	Patients are cared for in advance so that they do not fall ill at all
10	Previous effect	No idea
11	Principle of the "previously placed cushion"	A comprehensive template is being developed in which patients can very easily enter all the necessary documentation or refer to documents
12	Equipotentiality	No idea
13	Inversion (reversal of function)	It is not the patient who provides the documents, but the registration station obtains the documents itself
14	Sphericity (spheroidality)	No idea
15	Dynamization	The documents are not provided in one step, where everything is checked through, but piece by piece according to availability
16	Partial or excess effect	The patient is asked to simply provide everything, regardless of whether it is related to the current case or not
17	Transition to other dimensions	You don't ask the patient for documents but let him tell you everything he can think of in a conversation? (Biographical presentation)
18	Utilization of mechanical vibrations	No idea
19	Periodic effect	No idea
20	Continuity of the beneficial effect	If something is missing from the documentation, it is redone and money is made from it
21	Principle of rushing through (skipping)	No idea
22	Converting the harmful into the useful	No idea
23	Feedback	You go through the documents with the patient and ask additional questions to obtain further information
24	Principle of the intermediary	The patient is provided with a coach who translates from everyday language into hospital language
25	Self-service	The patient signs a power of attorney so that the hospital can take care of everything itself
26	Copy	All patient data is stored on a nationwide server from which the hospital can download the data
27	Cheap short life instead of expensive longevity	No idea

(Fortsetzung)

Table 1: (Fortsetzung)

No.	Innovation principle	Solution ideas
28	Replacing the mechanical system	No idea
29	Application of pneumatic/hydraulic systems	No idea
30	Use of flexible sleeves and thin films	No idea
31	Use of porous materials	No idea
32	Color change	No idea
33	Uniformity (homogeneity)	No idea
34	Removal and regeneration of parts	The registration process is completely separated from the hospital appointment
35	Change in physical and chemical properties	No idea
36	Application of phase transitions	No idea
37	Application of thermal expansion	No idea
38	Use of strong oxidizing agents	An oxidizing agent causes a strong reaction. The patient receives a reward, depending on the quality of the documents submitted
39	Application of an inert medium	No idea
40	Application of composite materials	Spouses, children, relatives, etc. are included in the collection of information

4 Discussion

This work was intended to show that TRIZ can also be applied to process management in hospitals and is not limited to the technical area. Very general examples were deliberately used in order to cover as wide an area as possible. Further evaluation and, if necessary, adaptation of the TRIZ system to the hospital environment is possible in order to validate the results and improve applicability of TRIZ in the hospital setting.

References

1. Hospital Association, D.: Press Release: Experts Paint Dramatic Picture (2023)
2. Flintrop, J.: Effects of the introduction of DRGs: economic logic becomes the measure of all things. Dtsch. Arztebl. Int. **103**(46), 3082 (2006)
3. Steffen, P., Offermanns, M.: Success-Critical Factors of Hospital Mergers. Research Report Commissioned by the Schubert Group and the Bank für Sozialwirtschaft. German Hospital Institute (2011)
4. Groß, C., Pfennig, R.: Digitalization in Industry, Trade and Logistics: Guide from Process Analysis to Application Optimization. Springer (2019)
5. Altshuller, G., And Suddenly the Inventor Appeared. Technical Innovation Center Inc. (2008)
6. Haines-Gadd, L.: TRIZ for Dummies. Wiley-Brand (2016)
7. Altshuller, G.: 40 Principles, TRIZ Keys to Technical Innovation. Technical Innovation Center Inc. (1994)

Funktionsanalyse und inkrementelle Verbesserung für das Vorgehen in der Funktionsanalyse

Oliver Gerundt(✉)

Robert Bosch GmbH, Gerlingen-Schillerhöhe, Deutschland
oliver.gerundt@de.bosch.com

Zusammenfassung. Das Grundbestreben bei TRIZ ist es, Dinge bzw. Systeme zu verbessern. Ein zentrales Werkzeug dabei ist die Funktionsanalyse, bei der ein System analysiert und bildlich dargestellt wird, um es anschließend zielgerichtet zu verbessern. Der Ablauf der Erstellung einer Funktionsanalyse ist sehr systematisch. Nach einer Komponentenanalyse erfolgt eine Interaktionsanalyse auf deren Basis eine tabellarische Funktionsanalyse erstellt wird. In der Praxis zeigt sich, dass der sinnvolle Schritt der Interaktionsanalyse leider häufig übersprungen wird, da sie in der späteren Analysephase nicht mehr verwendet wird. Die vorliegende Arbeit analysiert den Prozess der Erstellung einer Funktionsanalyse mittels einer Funktionsanalyse und leitet eine praktikable Verbesserungsmaßnahme ab, indem sie die Interaktionsanalyse mit der tabellarischen Funktionsanalyse verbindet.

Schlüsselwörters: TRIZ · Funktionsanalyse · Interaktionsanalyse · Trimming · X-Element

1 Ausgangssituation

Die Funktionsanalyse spielt in TRIZ eine zentrale Rolle. Sie schafft Systemverständnis und dient als ein wichtiger Baustein bei der Ableitung von Fragestellungen für die Verbesserung.

Gearbeitet wird in drei Schritten (s.a. [2] S. 29 ff.):

1. Komponentenanalyse
2. Interaktionsanalyse
3. Funktionsmodellierung
    3a. tabellarisches Funktionsmodell (Funktionstabelle)
    3b. graphisches Funktionsmodell

In der praktischen Anwendung ist häufig zu beobachten und zu erleben, dass die Schritte Interaktionsanalyse und die Funktionstabelle aus Zeitgründen nicht gemacht werden und direkt am graphischen Funktionsmodell gearbeitet wird. Dadurch werden Funktionen übersehen oder nicht ausreichend genau analysiert.

O. Mayer (Hrsg.): TRIZ-Anwendertag 2024,
https://doi.org/10.1007/978-3-662-70698-5_6

In den nächsten Schritten wird gezeigt, wie der Ablauf der Funktionsanalyse verbessert werden kann, indem Zeit eingespart wird, ohne die Vorteile einer Interaktionsanalyse und Funktionstabelle zu verlieren.

2 Funktionsanalyse für eine Funktionsanalyse

Um Ansatzpunkte für die Verbesserung der Funktionsanalyse zu finden, wurde für das Vorgehen zum Erstellen einer Funktionsanalyse selbst eine klassische TRIZ-Funktionsanalyse durchgeführt.

2.1 Die Komponentenanalyse

Der erste Schritt bei einer Funktionsanalyse ist die Analyse der Komponenten und der Supersystemkomponenten, den Komponenten außerhalb der Systemgrenze. ([2] S. 30 ff.)

In Abb. 1 ist die Komponentenanalyse dargestellt.

Technisches System:	Funktionsanalyse eines Systems
Zielkomponente:	Nutzer
Hauptfunktion:	Nutzer informieren
Zeit:	Erstellung der Funktionsanalyse

Systemkomponenten:	**Supersystemkomponenten:**
Komponentenanalyse	Nutzer
Interaktionsanalyse	System (das zu analysierende)
Funktionsanalyse (tabellarisch)	
Funktionsmodell (graphisch)	

Abb. 1 Komponentenanalyse einer Funktionsanalyse

Die Hauptfunktion einer Funktionsanalyse ist es den oder die Nutzer zu informieren, um das Wissen und die Erkenntnisse über das System zu erhöhen. Der Betrachtungszeitraum ist die Erstellungszeit einer Funktionsanalyse eines Systems.

2.2 Die Interaktionsanalyse

Der zweite Schritt der Funktionsanalyse ist die Interaktionsanalyse ([2] S. 43 ff.). Hier wird in einer Matrix ermittelt, welche Komponenten in Interaktion stehen. Bei mechanischen Systemen ist das im Normalfall ein physikalischer Kontakt, der über Berührung oder mittels eines Feldes entsteht. In unserem Fall eines organisatorischen Systems ist dieser Kontakt allgemeiner zu sehen. Folgende Matrix zeigt die Interaktionsanalyse für das System „Funktionsanalyse" (Abb. 2).

	Komponenten-analyse	Interaktions-analyse	Funktionsmodell (tabellarisch)	Funktionsmodell	Nutzer	System
Komponenten-analyse		X	-	-	X	X
Interaktions-analyse			X	-	X	X
Funktionsmodell (tabellarisch)				X	X	X
Funktionsmodell					X	X
Nutzer						X
System						

X → Interaktion
- → keine Interaktion

Abb. 2 Interaktionsanalyse einer Funktionsanalyse

Offensichtlich wird bei dieser Interaktionsanalyse, dass es sich um ein stark vernetztes System handelt, da nur drei Paare nicht in Interaktion stehen.

2.3 Das Funktionsmodell

Im nächsten Schritt wird aus der Interaktionsanalyse das tabellarische Funktionsmodell abgeleitet ([2] S. 60). Hierzu ist eine bewährte Vorgehensweise, zeilenweise durch die Interaktionsanalyse durchzugehen und für jedes Kreuz, das für Interaktion steht, eine oder mehrere Funktionen zu ermitteln. Es kommt auch vor, dass trotz Interaktion keine Funktion vorhanden ist, wenn am Objekt der Funktion kein Parameter geändert oder aktiv erhalten wird. Jede Funktion wird dann noch kategorisiert nach nützlich (useful, U) und schädlich (harmful, H). Die nützlichen Funktionen erhalten

Tab. 1 tabellarisches Funktionsmodell

Funktionsträger	Aktion (Verb)	Objekt der Funktion	Kategorie (U=useful, H=harmful)	Perform. level (I=insufficient, E=excessive, N=normal)
Komponentenanalyse	versorgt	Interaktionsanalyse	U	N
Nutzer	erstellt	Komponentenanalyse	U	N
System	informiert	Komponentenanalyse	U	N
Komponentenanalyse	beschäftigt	Nutzer	H	
Komponentenanalyse	informiert	Nutzer	U	N
Interaktionsanalyse	versorgt	Funktionsmodell (tabellarisch)	U	N
Interaktionsanalyse	beschäftigt	Nutzer	H	
Nutzer	erstellt	Interaktionsanalyse	U	N
System	informiert	Interaktionsanalyse	U	N
Interaktionsanalyse	informiert	Nutzer	U	N
Funktionsmodell (tabellarisch)	versorgt	Funktionsmodell	U	N
Nutzer	erstellt	Funktionsmodell (tabellarisch)	U	N
Funktionsmodell (tabellarisch)	beschäftigt	Nutzer	H	
System	informiert	Funktionsmodell (tabellarisch)	U	N
Funktionsmodell (tabellarisch)	informiert	Nutzer	U	N
Nutzer	erstellt	Funktionsmodell	U	N
Funktionsmodell	beschäftigt	Nutzer	H	
System	informiert	Funktionsmodell	U	N
System	informiert	Nutzer	U	N
Funktionsmodell	informiert	Nutzer	U	N

dann noch einen sogenannten Performancelevel, normal (normal, N), unzureichend (insufficient, I) oder exzessiv (excessive, E). Für die Funktionsanalyse ergibt sich damit folgende Tabelle (Tab. 1):

Mithilfe des tabellarischen Funktionsmodells wird das grafische Funktionsmodell abgeleitet, hier nur als Funktionsmodell bezeichnet. Daraus ergibt sich dann folgende Darstellung (Abb. 3).

Mit der Erstellung des Funktionsmodells ist ein wesentlicher Teil der Vorarbeit getan. Im folgenden Schritt werden die Fragestellungen für die Verbesserung des Systems abgeleitet.

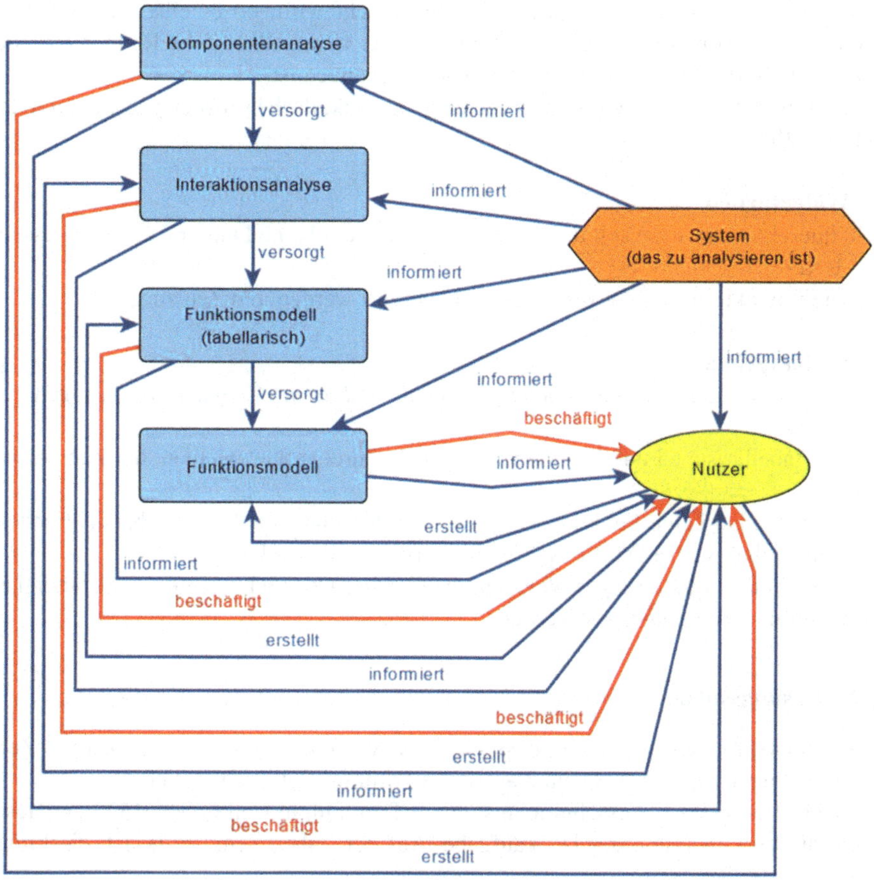

Abb. 3 Funktionsmodell einer Funktionsanalyse

3 Verbesserung des Systems

Auffällig im Modell ist, dass es viele schädliche Funktionen gibt, die auf den Nutzer wirken. Der Nutzer wird stark zeitlich eingebunden, weshalb in der Praxis häufig beobachtet werden kann, dass einzelne Schritte, insbesondere die Interaktionsanalyse und das tabellarische Funktionsmodell, nicht durchgeführt werden.

3.1 Herleitung der Aufgabe

Die Interaktionsanalyse und das tabellarische Funktionsmodell erzeugen Aufwand und werden später, wenn das grafische Funktionsmodell steht, meist nicht mehr genutzt. Trotzdem wirken sie unterstützend beim Erstellen des Funktionsmodells zu unterstützen. Die Interaktionsmatrix unterstützt, dass keine Interaktionen und damit auch keine Funktionen vergessen werden. Das tabellarische Funktionsmodell schafft nachvollziehbare Übersicht. Bei dessen Erstellung geschieht die eigentliche analytische Arbeit, bei der jede Interaktion untersucht wird, ob es eine Funktion gibt, mehrere Funktionen in die gleiche Richtung oder in gegensätzliche Richtungen oder ob eventuell trotz Interaktion keine Funktion vorhanden ist.

Daraus lassen sich die zwei folgenden „Physikalischen Widersprüche" ableiten ([1] S. 230 ff.):

1. **Widerspruch**
Die Interaktionsanalyse soll durchgeführt werden, um keine Funktionen zu vergessen
UND
die Interaktionsanalyse soll nicht durchgeführt werden, um Zeit zu sparen

2. **Widerspruch**
Das tabellarische Funktionsmodell soll erstellt werden, um Übersicht zu erzeugen
UND
das tabellarische Funktionsmodell soll nicht durchgeführt werden, um Zeit zu sparen

Es darf also keine Funktion vergessen werden und es soll eine Übersicht erstellt werden, wobei bei beiden Zielen Zeit eingespart werden soll

Daher die zentrale Frage: Wie kann gleichzeitig Übersicht erzeugt, Vollständigkeit gesichert und Zeit eingespart werden?

3.2 Lösungssuche

Den Widersprüchen ist gemein, dass zwei der Analyseschritte aus Gründen des Zeitaufwands unerwünscht sind. Ein klassisches Trimming ist hier möglich.

Abb. 4 zeigt zwei Ausschnitte aus einem Trimmingmodell ([1] S. 152 ff.). Dieses dient als Visualisierung der Denkaufgabe. Auf der linken Seite stellt sich die Frage,

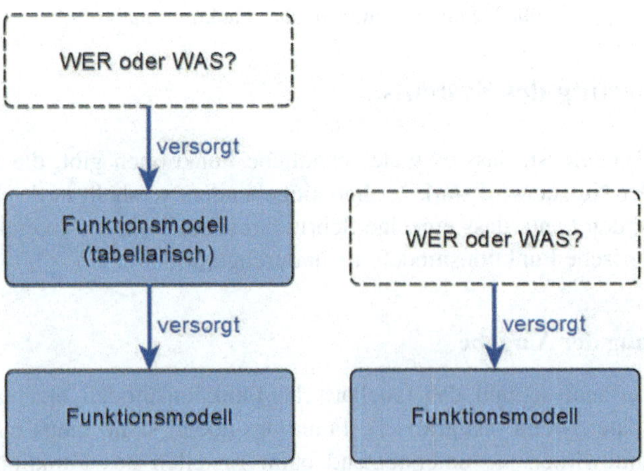

Abb. 4 Ausschnitt des Trimmingmodells

wie ein tabellarisches Funktionsmodell ohne Interaktionsanalyse erstellt werden kann. Auf der rechten Seite stellt sich die Frage, wie das (grafische) Funktionsmodell ohne tabellarisches erstellt werden kann.

Da es sich um zwei Komponenten mit unterschiedlichen Zielsetzungen handelt, die aufeinander aufbauen, wurde ein abstraktes Lösungsmodell mit der Einführung eines sogenannten X-Elements ([3] S. 47) erstellt. Dieses X-Element führt die benötigten Funktionen aus und erfüllt alle Anforderungen. Abb. 5 zeigt das gesamte Funktionsmodell mit X-Element.

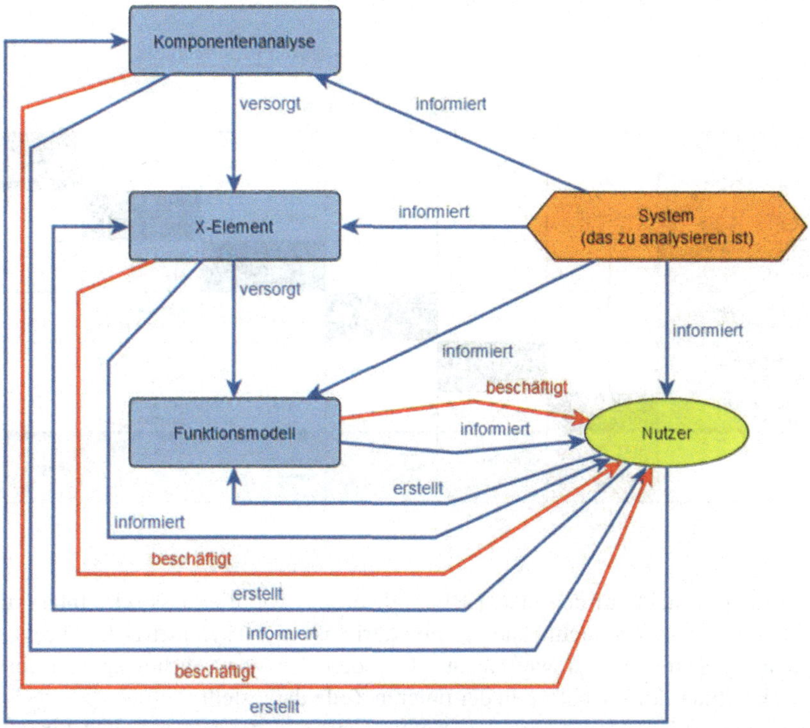

Abb. 5 Funktionsmodell mit X-Element

Unter Anlehnung an die unter 3.1 hergeleiteten Arbeitsfrage ergeben sich folgende Eigenschaften des X-Elements:

Das X-Element ermöglicht, dass keine Funktion vergessen, eine Übersicht erstellt und Zeit eingespart wird.

Für die Lösungsfindung eignen sich erfahrungsgemäß die Separationsprinzipien, auf die an dieser Stelle nicht genauer eingegangen wird.

3.3 Die Lösung

Im Rahmen der Suche nach einer abstrakten Lösung werden alle Separationsprinzipien der Reihe nach durchgegangen sowie die Ressourcen im System analysiert. Für die gegebene Fragestellung führte das Prinzip „Separation in der Zeit", zu einer Lösung. Bei diesem Prinzip werden u. a. die Zeiträume vor, nach und parallel zur operativen Zeit zur Lösung herangezogen. Betrachtet man die parallele Zeit, eine Art zeitlichen Verschachtelung (Innovationsprinzip 7), ist eine Lösung, die Interaktionsmatrix und das tabellarische Funktionsmodell gleichzeitig auszuführen. Hierzu wurden beide in ihrer Struktur kombiniert, weshalb der Name Kombimatrix gewählt wurde. Tab. 2 zeigt eine solche Matrix für das Beispiel „Funktionsanalyse".

Tab. 2 Kombimatrix

System	informiert (U,N)	informiert (U,N)	informiert (U,N)	informiert (U,N)	informiert (U,N)	
Nutzer	erstellt (U,N)	erstellt (U,N)	erstellt (U,N)	erstellt (U,N)		---
Funktions-modell	---	---	---		informiert (U,N) beschäftigt (H)	---
Funktionsmodell (tabellarisch)	---	---		versorgt (U,N)	informiert (U,N) beschäftigt (H)	---
Interaktions-analyse	---		versorgt (U,N)	---	informiert (U,N) beschäftigt (H)	---
Komponenten-analyse		versorgt (U,N)	---	---	informiert (U,N) beschäftigt (H)	---
	Komponenten-analyse	Interaktions-analyse	Funktions-modell (tabellarisch)	Funktions-modell	Nutzer	System

Die Matrix sieht auf den ersten Blick gleich aus wie die klassische Interaktionsmatrix, sie ist es aber nicht. Die Kombimatrix ist nicht symmetrisch! Die zu vergleichenden Komponenten wurden aus Gründen der Verwechslungsgefahr mit der klassischen Interaktionsanalyse in der unteren Zeile dargestellt.

Zu lesen ist die Matrix von den Komponenten in der ersten Spalte hin zu den Komponenten in der untersten Zeile. Die in der Tabelle markierte Funktion ist damit als „Nutzer erstellt Komponentenanalyse" zu lesen, eine „nützliche (useful)", „normale (normal)" Funktion,

Mit dieser Darstellung lässt sich ein Schritt in der Funktionsanalyse einsparen, Die übersichtliche Nachvollziehbarkeit ist gegeben und es erfolgt eine Kontrolle der Vollständigkeit der Funktionen.

Der neue Prozess zur Erstellung einer Funktionsanalyse stellt sich jetzt wie folgt dar:

1. Komponentenanalyse
2. Kombimatrix
3. Funktionsmodell (grafisch)

4 Zusammenfassung

Die Funktionsanalyse ist ein zentrales Element in TRIZ. Aus ihr werden Ansatzpunkte zur Verbesserung eines Systems abgeleitet. Damit ist es nur logisch, sie zu ihrer eigenen Verbesserung einzusetzen

Die gefundene Verbesserung ist kein revolutionär großer Schritt, sondern eher eine inkrementelle Verbesserung. Diese ist jedoch praxisgerecht und einfach anwendbar. Damit konnte eine Lösung für die zentrale Aufgabe („Wie kann gleichzeitig Übersicht erzeugt, Vollständigkeit gesichert und Zeit eingespart werden?“) gefunden werden.

Literatur

1. Adunka R.: *“TRIZ Level 1” Trainingsunterlagen;* TRIZ Consulting Group GmbH, Nürnberg (2014)
2. Ikovenko, Dr. S., Barkan, M. et.al: *“State-of-the-Art TRIZ, Theory of Inventive Problem Solving”*; Warsaw, Novismo, (2019)
3. Ivanov, G.: *“Formulas of Creativity”*; TRIZ Consulting Group, (2024)
4. Koltze, K., Souchkov, V.: “Systematische Innovation – TRIZ-Anwendung in der Produkt- und Prozessentwicklung”, 2 überarb Carl Hanser Verlag, München (2017)
5. VDI-Richtlinie 4521, Blatt 1: *„Erfinderisches Problemlösen mit TRIZ“,* Beuth Verlag, Berlin (2016)
6. VDI-Richtlinie 4521, Blatt 2: *„Erfinderisches Problemlösen mit TRIZ“,* Beuth Verlag, Berlin (2018)
7. VDI-Richtlinie 4521, Blatt 3: *„Erfinderisches Problemlösen mit TRIZ“,* Beuth Verlag, Berlin (2020)

Self-Functions in TRIZ: Between non-existing and omnipresent

Christian Iniotakis(✉)

Ulm University of Applied Sciences, Ulm, Deutschland
christian.iniotakis@thu.de

Abstract. The role of self-functions in TRIZ is rather ambivalent. Whereas the typical standard approach in Function Analysis regularly prevents them from even being identified, self-functions prominently appear in trimming techniques, are explicitly proposed in an Inventive Principle, and get involved in considerations about systems of ultimate ideality as well. This work points out some misunderstandings and inconsistencies in the standard perspective, elaborates the actual nature of self-functions and clarifies the relation to practical application. In addition, different trimming variants are shown to result in systems getting either cheaper or better, depending on the type of the trimmed function.

Keywords: TRIZ · Systematic Innovation · Function Analysis · Self-Function · Value Analysis · VA++ · Trimming

1 Introduction

Introduced by Gerasimov and Litvin, cf. [1], Function Analysis is one of the prime methods of modern TRIZ [2–6]. It allows to break down a given system into its components and to elaborate their functional interplay in a systematic, logical manner. The functional interaction between the system (or its components, respectively) and the system environment—referred to as supersystem in the following—is found out fully analogously. The result of Function Analysis serves as a basis for a variety of other TRIZ tools and methods, e.g. trimming considerations [5–8], systematic patent circumvention, or advanced value analyses with VA++ [9, 10]. The proper application of Function Analysis relies on specific TRIZ definitions of the terms 'component' and 'function', which are basic parts of standard TRIZ education and certification [4–6]. Thus, they are supposed to be common knowledge for regular TRIZ practitioners and not repeated here in detail for the sake of brevity. Nevertheless, for non-TRIZ practitioners it should be pointed out, that the notion 'component' does not only cover parts of the system, but also the system itself, as well as the supersystem and its parts. Generally, all objects of mass are defined to be TRIZ components and can conceptually be combined/grouped or decomposed/broken down into other components. A very special case of a TRIZ function is a so-called self-function, which reflects the situation

O. Mayer (Hrsg.): TRIZ-Anwendertag 2024,
https://doi.org/10.1007/978-3-662-70698-5_7

of a TRIZ component changing or controlling itself. The occurrence of self-functions is rather selective in TRIZ. On the one hand, a self-function is commonly known as the standard solution pattern for trimming rule B [5–8] and explicitly relates to Inventive Principle #25 "Self-Service", cf. e.g. [2, 3, 5, 6, 8]. Furthermore, there is some connection to those self-functions described by Mann to be considered for approaching or envisioning systems of ultimate ideality [2]. On the other hand, applying Function Analysis according to standards systematically prevents self-functions from being identified, since the common practice is to mark the diagonal in the interaction matrix with a negative entry by default, cf. e.g. [11], or to completely neglect the diagonal right away, cf. e.g. [5, 6]. Thus, a deeper consideration about the existence of self-functions is regularly skipped in the decisive and final part of Function Analysis: Eventually all the relevant functions are modelled exclusively out of positive entries in the interaction matrix only. As a direct consequence of this systematic neglect of self-functions there is an obvious formal inconsistency in using TRIZ methods in combination: Assume, a system is analyzed with Function Analysis, then trimming rule B is successfully applied, and finally the resulting system is re-examined with Function Analysis again from scratch for any reason. Just following the standard procedures, the second Function Analysis will certainly miss the very self-function, which was elaborated in the intermediate trimming step, and which is so crucial for the improved, leaner system to properly work and operate. This current standard or standard behavior of systematically filtering-out self-interactions and self-functions during Function Analysis seems to be so clear and is supposed to be so trivial, that typical training material does not even give any argumentation or explanation about it at all, cf. e.g. [5, 6]. In publication [11], Adunka mentions that "a component could interact with itself if the relevant field is not modeled" to conclude "this is very seldom the case". Communication with TRIZ practitioners—from all levels of experience—reveals more or less the same opinion: The overwhelming majority of the TRIZ community seems to foster the trivial understanding of self-functions being non-existent in nearly all settings, and/or being irrelevant in the rare exceptional cases anyway. However, both common attributions to TRIZ self-functions, i.e. non-existence and irrelevance, are generally false as demonstrated in the following.

2 System Perspectives

There also is a more fundamental, conceptional approach to self-functions. From the system perspective, every internal function, e.g. a regular auxiliary function between two system components, fulfills all requirements to be identified as a self-function of the system itself. Obviously, it is a TRIZ function, performed by (some part of) the system on (some part of) the system. The same is true for any self-function of a system component: It is a self-function of the system as well. Just for a better understanding, this is illustrated for some simple case in Fig. 1. Consequently, any internal function from any sub-system level turns out to be a self-function of the system itself. From this perspective it becomes clear, that systems such as a car or a computer carry tremendous amounts of self-functions, because they involve a lot of relevant internal functions on all kinds of sub-system levels which are necessary for regular operation.

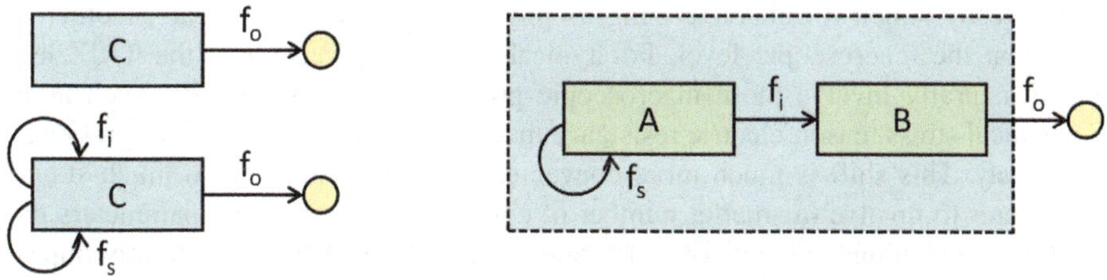

Fig. 1. (Left, top) A component C performs a function f_o on some other component, which is of no deeper interested here and therefore indicated by a circle only. (Right) Just for illustration, component C is assumed to consist of two (sub-)components A and B, providing two relevant internal functions: A regular auxiliary function f_i and a self-function f_s of component A. (Left, bottom) Since both internal functions f_i and f_s also fulfill the definition of a self-function of component C and are identified as relevant for its internal functional interplay, they could potentially be displayed as self-functions on the higher level as well, as an alternative.

It is important to understand, that this kind of system perspective does not only apply to complicated, cleverly engineered or designed systems, but to any macroscopic component, even simplest objects. For example, an ordinary screw performing a holding function virtually consists of smaller material parts holding each other; bringing this scale even down to the microscopic level reveals zillions of atoms making up the screw and mutually holding together. All these atomic holding functions are also self-functions of the screw. And in sum, they are extremely relevant for its macroscopic functional performance: Only because of them, the screw holds itself in its shape, and is stable enough to do its job of holding something else.

3 Practice and Parameters

As demonstrated in the previous section, typical components and/or systems generally carry an abundance of self-functions from various sub-levels, all of which are relevant, which is quite in contrast to standard opinions and explanations. But from the pure pragmatic perspective, two noteworthy aspects need to be addressed. Firstly, identifying, elaborating, or even simply displaying all those self-functions could easily result in a methodical overkill in practice. Secondly, the current approach of fully ignoring those self-functions in a standard Function Analysis does not really seem to have a significant negative impact in many practical application settings. It turns out, that both aspects can be explained by the same insight, based on the system structure being composed of different nested system levels or scales. In the interdisciplinary field of complex systems, it is basic knowledge, that interactions on lower system levels are certainly fundamental for the—sometimes surprising—system behavior itself; but for actually describing, modeling, or understanding it, macroscopic features of the system level are regularly supposed to be the better choice, cf. e.g. [12, 13]. Those self-functions of a component representing internal functions and interactions on an atomic level, for example, also are the intrinsic origin of component properties perfectly suited for characterizing its physical, chemical, technical, or other behavior on the macroscopic level. For typical application scenarios, the TRIZ team would naturally involve these macroscopic properties and parameters—such as the mechanical stress tensor, electric resistance, heat conductivity, viscosity, or pressure—right away. This shift is much more convenient as well, since it is so much simpler and clearer to involve (a smaller number of condensed) macroscopic parameters than handling tremendous amounts of self-functions, a task supposed to be totally impractical if not impossible in typical settings. Thus, self-functions originating from a lower level should equivalently, effectively, and conveniently be treated as parameters, properties or features at the higher level as often as possible. This approach does not only hold for typical material parameters of simple components, as mentioned above, but also for the top-level description of systems from the customer perspective with the help of MPVs, cf. e.g. [6]. For example, the performance of a computer system—in the sense of how fast it is in executing given tasks—is a top-level system parameter effectively summarizing the combination of a variety of system internal functions, all of which are self-functions of the computer system itself. Thus, in practice, the prime essence of many relevant self-functions is typically included in the considerations of the TRIZ team, even though they are not found in Function Analysis and the team might not even be aware of their existence—as long as the team properly knows about important parameters and features of the components, as well as the resources provided by them. However, it is still recommended to explicitly check for remaining self-functions of importance, e.g. by simply marking the diagonal of the interaction matrix with a positive entry by default. Otherwise—following the standard procedure—important insight and useful information about the components or the system might be lost or overlooked. This is particularly true, if the basic features associated with a self-function are none of the classical parameters, and/or the team might not be aware of its deeper connection to relevant aspects of functionality. For example, screws combined with some specific counterparts serve as self-adjusting connectors in a larger furniture system. Standard Function Analysis will certainly reveal the connectors to hold parts, without any further need to elaborate the actual origin from internal holding functions on the microscopic level. In contrast, the feature of self-adjusting, which is due to direct internal functions of the connector, namely between the screw and the counterpart, is likely to be overlooked, although it is relevant for several important aspects of the furniture system, e.g. the time for setting it up, and the accuracy of the final arrangement. Thus, this particular insight might prove quite valuable for advanced analyses involving costs, functionality, and ideality, for example, and its loss can be avoided in a straightforward way by simply checking each component for additional self-functions considered to have the most important impact on the system.

4 Functional Value and VA++

In TRIZ application scenarios focusing on functionality, ideality, or value, respectively, omitting explicit self-functions is not at all problematic in some sense. In the VA++ framework as discussed and derived in [9, 10] in detail, the functional value

δ of any TRIZ component with respect to a given functional aspect is simply determined by its net functional contribution. This is the sum of all functional values of its performed functions, minus that of all its received functions. Since self-functions equally count to both functional flow directions, their contributions always cancel out each other, and they formally drop out in this calculation [10]. Even though self-functions are extremely important from the causal, functional perspective, e.g. for internally enabling a component to perform functions on other components, they do not need to be explicitly involved in the sense of functional value: Their impact already is and must be fully accounted for in the functional value of exactly the non-self-functions, which really represent the functional exchange with other, different components. For example, as mentioned above, the holding function of the screw is only possible due to many internal holding functions. And the functional value assigned to this macroscopic effect or ability of the screw already fully reflects the aggregated contributions of all the internal interactions and functions on or from lower system levels.

5 Trimming Aspects

As mentioned in the introduction, a prominent appearance of self-functions in TRIZ is found in a successful application of trimming rule B [5–8]. A mathematical derivation reveals that the functional value of the component, which is enabled to additionally perform a useful function on itself, increases accordingly—in exact correspondence to the transfer of the trimmed function [10]. Based on these key findings of VA++, there is a substantial difference between successful trimming scenarios, depending on the origin of the useful function to be trimmed. Note, that secondary effects of a successful trimming, such as additional implementation or component costs, are equally neglected in the following for the sake of simplicity. In a first and rather common scenario, a system component successfully takes over an internal auxiliary function, as illustrated in Fig. 2, a) and b). As a result, the system can become leaner and cheaper in case some other component is rendered unnecessary, but the functionality of the system remains unchanged. In contrast, if the trimmed useful function taken over by a system component originates from the supersystem, the system does not get leaner or cheaper, but better, cf. Fig. 2, c). Effectively, the system requires less functional input, and its own functionality, measured in form of the functional value δ, increases accordingly [10]. Even though functional input is directly connected to costs in some cases, there is a general difference in a system becoming cheaper or better—in particular with respect to customer perspectives and competition. Accordingly, the Inventive Principle #25, "Self-Service", cf. e.g. [2, 3, 5, 6, 8], can be understood to primarily point in the direction of a better system. Regarding the self-functions described by Mann [2, Chap. 19], it should be pointed out, that there the notion of self-function is not always the same as used in this work. Here, a self-function is a function both performed and received by the identical TRIZ component. With respect to trimming, the standard approach is to enable the component or system to provide a useful self-required function now by itself, analogous to rule B and the scenarios in Fig. 2.

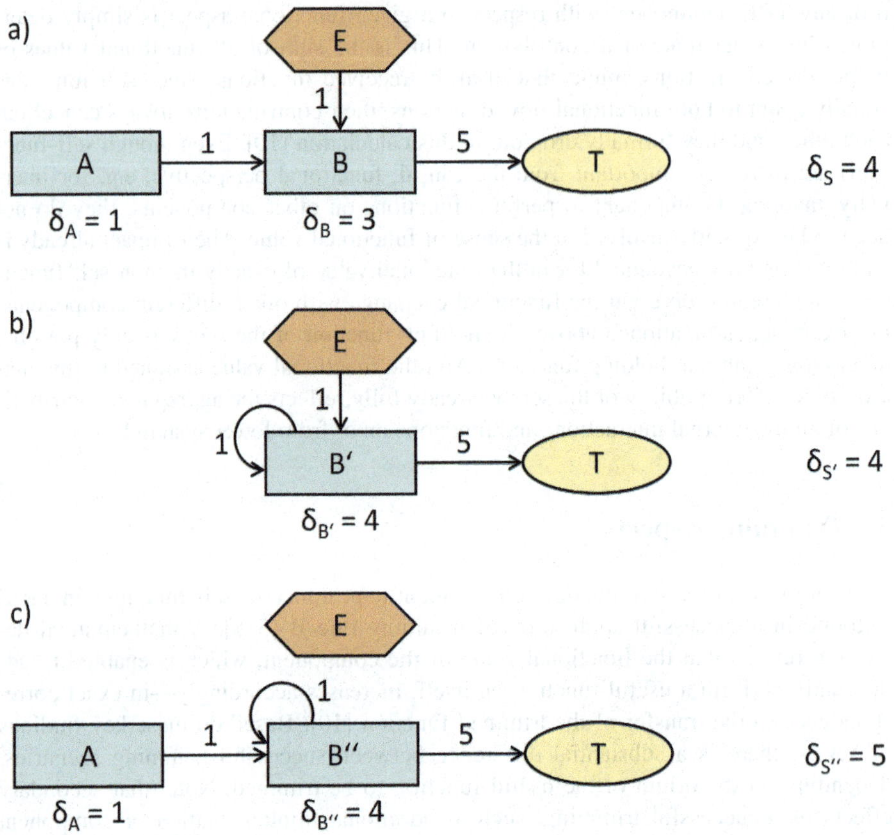

Fig. 2. **a** An original system S of two components A and B provides a main function to the target component T. It requires functional input from the environmental super system component E as well. The given functional values of the functions are for illustrative purposes only. The functional values δ of system components and the system itself are measures for their functionality and are simply determined by the net functional contribution according to VA++. **b** An upgraded system S' is developed, in which an upgraded component B' additionally takes over the former internal auxiliary function of A itself. The system becomes leaner and cheaper, but its functionality remains the same. **c** Based on S, some other upgraded system S'' is developed, in which an upgraded component B'' additionally takes over the external input function—originally provided by E—itself. The system does not get cheaper, but better due to its increased functionality as indicated by the functional value.

In some other context, however, a self-function only refers to the pattern of a component or system now doing something useful by itself, but not necessarily also on itself.

For example, a so-called self-drilling screw does not drill itself as an object, but acts on something else, e.g. thin metal sheets. What the screw actually does, is to perform the drilling job as such, which was previously provided by some other means, now by itself. In terms of trimming, this corresponds to rule C. It is obvious to see, that also these types of self-functions represent the approach of making the system better: Additionally acquiring other useful functions acting on the supersystem, e.g. the target, simply aggregates more functionality.

6 Resume

This work shows that self-functions are neither rare nor unimportant. Rather, it's the exact opposite: Even simplest TRIZ components exhibit an abundance of relevant self-functions, that manifest themselves in macroscopic parameters and features of the component. In practice, it is recommended to explicitly check for further self-functions of importance for the specific problem setting by putting positive entries in the diagonal of the interaction matrix as a default. Those self-functions could, for example, originate from internal auxiliary functions of the component itself. Application of the VA++ framework reveals that a successful trimming operation according to rule B—with a system component performing a required useful function by itself—results in different system changes, depending on the type of the trimmed function. The trimming of an internal auxiliary function acts towards a leaner and cheaper system, whereas the trimming of an external input function makes the system better.

References

1. Litvin, S., Feygenson, N., Feygenson, O.: Advanced function approach. Procedia Eng. **9**, 92–102 (2011). https://doi.org/10.1016/j.proeng.2011.03.103
2. Mann, D.: Hands-On Systematic Innovation. 2nd edn. IFR Press (2010)
3. Gadd, K.: TRIZ for Engineers: Enabling Inventive Problem Solving. Wiley (2011)
4. VDI 4521-2. Inventive problem solving with TRIZ. VDI (2018)
5. Litvin, S., Ikovenko, S., Lyubomirskiy, A., Stevenson, K.: MATRIZ-Level 1 Training Manual. GEN-TRIZ (2019)
6. GEN TRIZ Knowledge Transfer: Basic Module, Manual. GEN-TRIZ (2019)
7. Sheu, D.D., Hou, C.T.: TRIZ-based systematic device trimming: theory and application. Procedia Eng. **131**, 237–258 (2015). https://doi.org/10.1016/j.proeng.2015.12.385
8. VDI 4521-3. Inventive problem solving with TRIZ. VDI (2020)
9. Iniotakis, C.: VA++—the next generation of value analysis in TRIZ. In: Systematic Innovation Partnerships with Artificial Intelligence and Information Technology. TFC 2022. IFIP Advances in Information and Communication Technology 655, 29–38 (2022). https://doi.org/10.1007/978-3-031-17288-5_3
10. Iniotakis, C.: Functional value of TRIZ components—universal concept and direct implications. In: Systematic Innovation Partnerships with Artificial Intelligence and Information Technology. TFC 2023. IFIP Advances in Information and Communication Technology 682, 301–316 (2023). https://doi.org/10.1007/978-3-031-42532-5_24
11. Adunka, R.: Function analysis for electronic products. In: Proceedings of the TRIZ Future Conference 2010, pp. 165–171. Bergamo University Press (2010)
12. Thurner, S., Hanel, R., Klimek, P.: Introduction to the Theory of Complex Systems. Oxford University Press (2018)
13. Sayama, H.: Introduction to the Modeling and Analysis of Complex Systems. Open SUNY Textbooks (2015)

AI for TRIZ: ChatGPT vs. Kangaroo

Christian Iniotakis(✉) und Volker Herbort

Ulm University of Applied Sciences, Ulm, Deutschland
{christian.iniotakis,volker.herbort}@thu.de

Abstract. This study investigates, if capabilities of text-based AI tools are already able to support the systematic innovation method TRIZ by encompassing the three important elements of analysis, ideation, and evaluation in one go. For this purpose, ChatGPT 3.5—as a prime, free and famous representative—is tested on a sample set of 50 Kangaroo questions. After presenting the overall results of the statistical evaluation the focus is set on detailed aspects and particular issues of some selected questions and answers. Apart from different kinds of answer distributions this also covers exemplary sets of observed inconsistencies and effects such as hallucinations, indicating underlying intrinsic problems. As a key result, the current performance is not satisfying, yet.

Keywords: TRIZ · Systematic Innovation · AI · ChatGPT · Chatbot · Kangaroo · Mathematics · LLMs

1 Introduction

The new opportunities offered by artificial intelligence are omnipresent and manifold, also with respect to innovation activities of all kinds, cf. e.g. [1]. Particularly AI tools based on large language models (LLMs) are increasingly involved in various types of applications, but still facing challenges, cf. e.g. [2]. The potential impact of such a technology is also discussed in the field of the systematic innovation method TRIZ, cf. e.g. [3]. In practice, TRIZ application in the industrial context ranges from days to years, and from simple setups to more sophisticated processes. In any case, however, for its goal of innovation, improvement, and/or problem solving TRIZ regularly contains three important elements, namely analysis, idea generation, and evaluation. The element of analysis is about really understanding the sometimes very complex problem setting and/or the specific situation in general. The element of idea generation is about generating and collecting all kinds of ideas and naturally involves creativity. The third element, evaluation, typically relies on subsequent studies, feasibility tests, taking into account valid restrictions, etc., i.e. cross-checking with reality and facts. Even though all these three elements are supported by various systematic innovation methods, still it's basically about understanding, creativity, and factual correctness. That AI tools offer a great potential in the one and/or the other element is without question. The key intention of this work, however, is to test, if a freely available AI

O. Mayer (Hrsg.): TRIZ-Anwendertag 2024,
https://doi.org/10.1007/978-3-662-70698-5_8

technology based on LLMs, as provided by ChatGPT 3.5 (ChatGPT) [4], has some potential to support all three of those required elements simultaneously, in one go. In other words, this study is not about AI tools being applied in some stages or methods of TRIZ or partially supporting human TRIZ professionals in the one or other aspect. Rather, it addresses the much more fundamental question, if or to which extent such an AI technology is already capable of solving TRIZ-like problems fully autonomously, as a stand-alone tool. Accordingly, the testing setup needs to cover all three core elements of TRIZ in quite a compact way. Therefore, Kangaroo questions are considered to provide a suitable basis for testing.

2 Kangaroo

The term 'Kangaroo' stands for a mathematics competition for pupils—also referred to as students in the following -, which started in Australia in 1978 and grew very international in the following years [5, 6]. In 2023, for example, the competition took place in over 90 countries around the globe and had over 5.7 million participants in total, coming closer to the pre-pandemic level of 6.2 million participating pupils in 2019 [7, 8]. On the level of individual countries, the participation numbers are quite different, as can be seen by comparing, e.g. Brasil (> 900.000) and Germany (> 800.000) with the US (< 40.000) [7, 8]. Even though the Kangaroo competition is based on the same annual test preferably to be scheduled at the very same day for all participating countries, its prime intention is neither competition nor selection. Rather, "[t]he game contest aims at attracting as many students as possible without conducting any national selection or comparison between countries" [9]. Accordingly, the Kangaroo test is designed to promote the joy for mathematics and to encourage analytic work, logical thinking as well as the creative elaboration of riddles and smaller problem settings. Typical Kangaroo questions do not just refer to specific textbook knowledge but allow for several different ways and approaches to lead to the unique and correct answer, involving creativity, smartness, or common sense [8]. Thus, their solution regularly contains and/or requires all three of the core elements of systematic innovation, from really understanding the question, to generating ideas how to come to an answer, which eventually is elaborated and hopefully evaluated to be correct. Formally, Kangaroo questions are multiple choice, and there are six different test levels depending on the pupils age—or grade, respectively. The first level is related to pupils of approximately 6–8 years, whereas the sixth level is for participants of roughly 17–19 years.

3 Setup

A total of 50 test questions is extracted from sets of Kangaroo sample questions provided by Math Kangaroo USA, which are publicly available in the web together with the correct answer [10]. Per year, one such set of 18 sample questions is published, with three sample questions (a simple, a medium, and a more difficult one) for each of the six Kangaroo levels. Here and in the following these six levels are represented by

US grades ranging from 1 & 2 up to 11 & 12. The published sets of the years 2011–2023 form a basis of 234 Kangaroo sample questions in total.

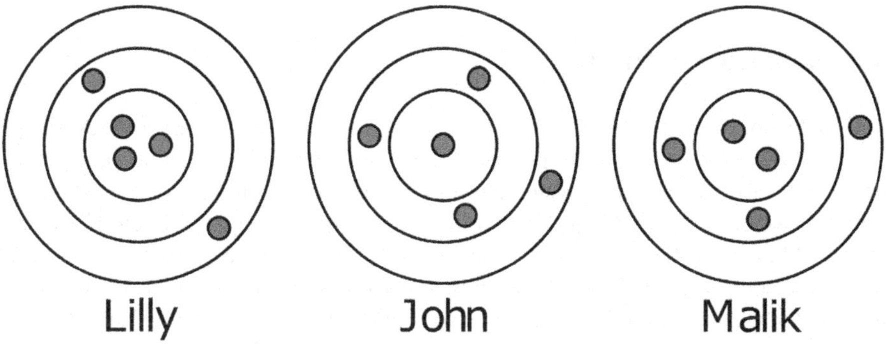

Fig. 1. Kangaroo test questions regularly involve visualizations and graphics in various ways. This example taken from [8] illustrates just one of these cases, with the graphics containing necessary information for the correct solution, without being explicitly referred to in the question text: "Lilly, John and Malik each throw five arrows at a target. Arrows that land in the same ring give the same score. Lilly has scored 46 points and John 34 points. How many points did Malik score?" The points from 37 to 41 are given as the five multiple-choice options [8].

From these, those test questions are filtered out, which seem to be suitable for a most fair testing of text-based tools such as ChatGPT. For example, typical Kangaroo questions require the understanding of associated graphics or illustrations for finding a correct solution as depicted in Fig. 1, which is beyond the scope of AI capabilities to be tested in this study. Reasons for discarding a given sample question are, e.g.:

- It relies on a graphics or figure in any way or contains a textual reference to it.
- It involves some more specific mathematical formula notation such as 2^{2021}.
- It relies on specific information, which has changed since then or is mistakable, such as the "current year".
- It relies on the knowledge of the answering options for a correct result or textually refers to the answering options in any way, instead of e.g. just asking for the result.

However, a sample question accompanied by an illustration, which is neither needed for deriving the correct answer nor referenced to in the text, qualifies for testing. This filtering process eventually results in the final set of 50 questions considered suitable for testing. It should be noted that these 50 test questions are not evenly distributed over the six Kangaroo test levels. For grades 5 & 6 as well as for grades 11 & 12 there is only the minimum number of six test questions, while the level related to grades 7 & 8 contains the maximum number of eleven ones. In October and November 2023, all the 50 test questions have been fed to ChatGPT for five individual trials each, resulting in 250 single test trials in total. For seven of those 50 questions the analysis is extended to even 20 trials each. During the testing process, the same question is not asked again within the very same chat, and feedback to an answer is never provided to minimize ChatGPT's answering behavior being influenced by history or interaction.

No specific prompting or advice apart from the pure question text are involved, such that there is no direct need to consider additional interface effects. In particular, the multiple-choice options as such are not transferred to ChatGPT for the same reason.

4 Statistical Overview

In a first step, the total of 250 answers—five for each of the 50 Kangaroo test questions—provided by ChatGPT are processed manually and evaluated to be either wrong or correct. Note, that every given answer contains a clear final statement in the last sentence which is solely taken for this kind of evaluation. Thus, a ChatGPT answer is counted as correct as soon as the final statement declares the correct solution, even if this correct result involves several wrong steps and/or sheer luck, as elaborated in more detail in the next section of this study. Final statements not providing the correct answer, but falsely claiming that more information is necessary for obtaining the solution, that no solution exists at all, or similar are counted as wrong. Such statements are observed in roughly 10% of the answers and seem to be correlated to specific questions. In several cases, there are final statements such as "at least 12" (correct answer: 11) or "47 or 32" (correct answer: 31), which are obviously wrong as well. The result of this evaluation is depicted in Fig. 2. All 250 single trials are displayed

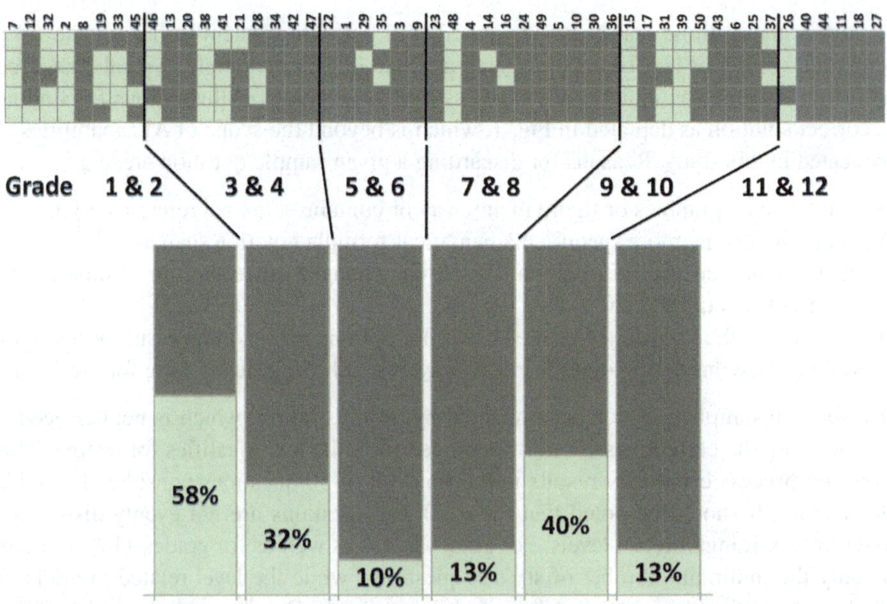

Fig. 2. (Top) Evaluation of ChatGPT's answers to the 50 Kangaroo sample questions (arranged horizontally) with 5 trials per question (arranged vertically in an upwards direction) Bright color represents correct answer(s). ID-numbers of the actual questions are indicated at the very top. (Bottom) Resulting percentage of correct trials per grade level.

in the upper part of the graphics, while the lower part shows the final statistical result. Correct answers are indicated in bright color, wrong answers in dark. To clarify the information contained in the figure, the intermediate Kangaroo level of grade 5 & 6 is represented by the six test questions with ID-numbers 22, 1, 29, 35, 3, and 9.

Table 1. Ratio of Kangaroo test questions reliably answered correctly by ChatGPT in all relevant trials, for the total set of 50 test question and within each particular grade level. First data line: Result after the five standard trials per question. Second data line: Including the extended analysis of 20 trials for those six questions supposed to be reliable after the first five trials, as shown in detail in Fig. 3

TOTAL	1 & 2	3 & 4	5 & 6	7 & 8	9 & 10	11 & 12
12% (6/50)	**25%** (2/8)	**10%** (1/10)	**0%** (0/6)	**9%** (1/11)	**22%** (2/9)	**0%** (0/6)
4% (2/50)	**13%** (1/8)	**0%** (0/10)	**0%** (0/6)	**0%** (0/11)	**11%** (1/9)	**0%** (0/6)

ChatGPT's answer to the fourth trial of question 29 was correct, as well as the third and fifth trial of question 35. All other trials in that grade level are evaluated to be incorrect. In sum, 3 out of 30 trials are correct in that particular grade, resulting in 10% overall correctness. The statistical result basically indicates, that ChatGPT performs much better for lower grades, i.e. predominantly on those test questions which are considered simpler for human participants and/or suited for younger pupils. However, an important aspect is, that the original decision, which difficulty, or grade to attach to a given Kangaroo question naturally involves some tolerance and simplifies a variety of different difficulty aspects to just one grade. Moreover, the assumed difficulty level depends on the actual circumstances as well. For example, pupils are not allowed to use any tools apart from pens in the Kangaroo competition, so they can neither rely on pocket calculators nor math books. Thus, questions considered somehow difficult for students in such a situation would certainly be estimated much simpler if computer access was allowed during the contest. This might partially explain why ChatGPT performs better for some of the more 'difficult' questions in higher grades. In grade 9 & 10, for example, the correct solutions of questions 15 and 39 just require a single multiplication or division, respectively. In the first case, it involves minutes and seconds, in the second case larger numbers, so it might not be considered too simple for regular human participants, while it really is trivial for any kind of computer-based tools. These two questions alone make up for more than half of ChatGPT's 40% success rate in that grade. Another example is question 48 in grade 7 & 8. Here, the correct solution is straightforwardly achieved out of the numbers of sides of cubes and tetrahedrons. Whereas the factual knowledge of a cube having six sides certainly is available for typical human participants, the decisive part is about the tetrahedron. If the student knows this particular geometric shape, the question is rather trivial to answer; if he or she doesn't, or can't associate the name to the shape, it is far more 'difficult' in the sense of being even impossible without guessing.

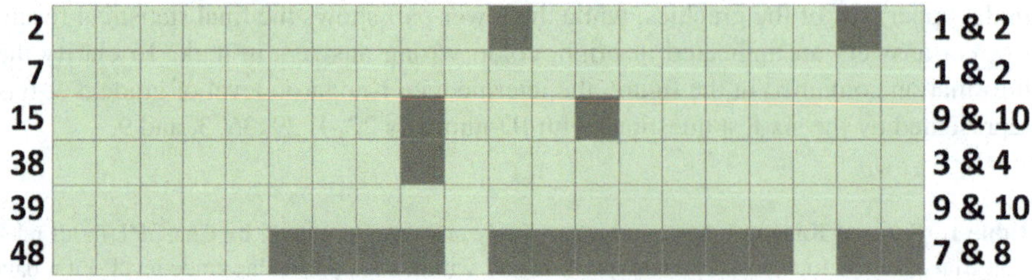

Fig. 3. Extended analysis of exactly those six Kangaroo test questions answered correctly by ChatGPT in all first five standard trials. For each question, 15 more trials are performed, and the result of all 20 trials is arranged horizontally from left to right, with bright color indicating correct answer(s). The numbers on the left denote the question ID, the numbers on the right the grade level. Obviously, errors may occur every now and then, unexpectedly, or even in larger sequences.

Therefore, any tool with access to a larger knowledge base should not struggle here, and neither does ChatGPT—at least not yet in these five trials. A second major aspect indicated by the results is about repeatability or reproducibility: For around two thirds of the questions, ChatGPT gets it fully right or fully wrong within the five trials, but the remaining third shows some ambiguity. Such an ambiguity can be found for any grade level. Since this study aims to test an AI tool also reliably covering the element of correctness itself, not relying on a subsequent control, correction, or statistical evaluation step by e.g. a human user, a key requirement is to generate the correct answer not only every now and then, but literally always. Taking that perspective into account, the performance of ChatGPT is much worse, as seen in the first data line of Table 1. In sum, only six of 50 questions are answered correctly for five times in a row, representing 12% of the selected sample set. Two of these questions belong to grade 1 & 2, and one to grade 3 & 4. The other three are given by questions 15 and 39 of grade level 9 & 10, as well as question 48 of grade level 7 & 8, all of which are already addressed above in the context of presumably being simpler for AI than for human participants. For checking the reliability in more detail an extended analysis is performed on exactly these six questions by adding 15 more individual trials. The result of those 20 trials on each of the six most reliably answered questions is depicted in Fig. 3. As a key result, 5 trials are obviously not enough to get a somehow meaningful, reliable answer about repeatability. Only two of the six questions happen to be answered correctly by ChatGPT also in the 15 extra trials. Three of them seem to be answered incorrectly every now and then. For one question—number 48, the one asking for sides of cubes and tetrahedrons -, ChatGPT even generates larger sequences of wrong answers. Some more detailed analysis about the specific content in these blocks is presented in the next section of this work. The strict requirement for reliable, repeatable factual correctness drastically limits ChatGPT's performance: As seen in the second data line of Table 1, there are only two of the 50 Kangaroo test questions ChatGPT correctly answers in all performed trials within the scope of this study—despite the pre-selection of the sample set for being most suitable. Moreover, the resulting 4% success rate can only be regarded as the upper bound based on 20 trials. A larger statistical approach might impair this number even more.

5 Detailed Aspects

Human readers of a text that is perceived as reasonable naturally assume a corresponding understanding on the part of the author. It is important to note, that answers generated by ChatGPT can of course be reasonable in that sense as well, at least to some extent or in some parts or patterns. After all, such tools have been heavily trained on original texts presumably containing an abundance of human understanding. On the other hand, generating an answer basically is a pure statistical process. In any case, ChatGPT is not supposed to have its own 'real' understanding. Nevertheless, it is quite interesting to explore the outcome of such a complex interplay between original understanding and statistical intermixing.

5.1 Answer Distributions

As explicitly mentioned and illustrated in the previous section, ChatGPT is tested for 20 independent trials on the very same question 48: "Ellie has 5 cubes and 3 tetrahedrons. How many sides do these solids have all together?" Since a cube has six and a tetrahedron four sides, the correct answer is straightforwardly achieved by multiplication and addition, resulting in 42 sides altogether. It turns out, that ChatGPT generates the correct answer in 35% of the cases, and—within the given number of trials—the answer distribution is restricted to three different values: Apart from the correct value 42, the wrong values 156 and 29 occur as well. The full distribution is depicted in Fig. 4. To explore the origin of the two wrong values, some typical answers are given in the following. Here and in the rest of the work, the original formatting of ChatGPT output may be compactified for the sake of brevity, sometimes requiring the addition of signs such as a period as indicated. The text itself is unchanged. Highlighting is made by the authors.

ChatGPT #19:

"To find the total number of sides for the cubes and tetrahedrons combined, you need to sum the number of sides for each type of solid. A cube has 6 faces, and each face is a square, so it has 4 sides. **Therefore, each cube has 6 × 4 = 24 sides.** A tetrahedron has 4 faces, and each face is a triangle, so it has 3 sides. **Therefore, each tetrahedron has 4 × 3 = 12 sides.** Now, for the given quantities: Ellie has 5 cubes, so the total number of sides for the cubes is 5 × 24 = 120 sides. Ellie has 3 tetrahedrons, so the total number of sides for the tetrahedrons is 3 × 12 = 36 sides. To find the overall total, add the number of sides for cubes and tetrahedrons: Total sides = Sides of cubes + Sides of tetrahedrons[.] Total sides = 120 + 36 = 156[.] Therefore, the cubes and tetrahedrons together have 156 sides."

ChatGPT #12:

"To find the total number of sides for the cubes and tetrahedrons, you need to calculate the sum of the number of sides for each type of solid. A cube has 6 faces, and each face is a square, so it has 4 sides. Therefore, the total number of sides for the cubes is 5 cubes × **4 sides/cube** = 20 sides. A tetrahedron has 4 faces, and each face is a triangle, so it has 3 sides. Therefore, the total number of sides for the tetrahedrons is 3 tetrahedrons ×

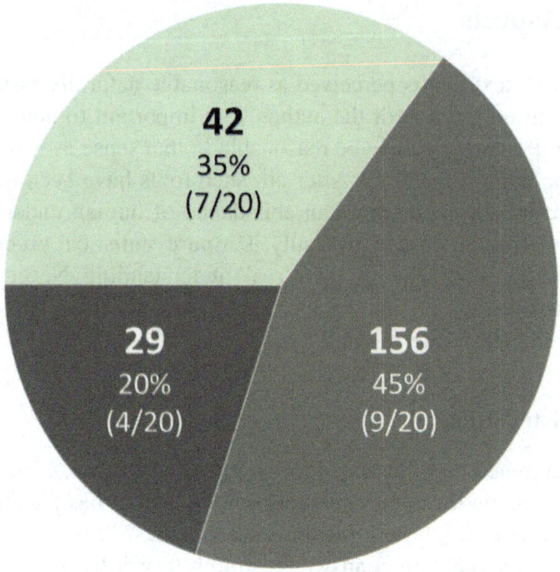

Fig. 4. ChatGPT's answer distribution for 20 independent trials on Kangaroo question 48 consists of three different values only. The correct value 42 is indicated by the bright color.

> **3 sides/tetrahedron** = 9 sides. Now, add the sides of cubes and tetrahedrons together: 20 sides (cubes) + 9 sides (tetrahedrons) = 29 sides[.] So, all together, the cubes and tetrahedrons have 29 sides."

Apparently, the generation of both wrong values seems to be based on the strong correlations of a dice to a square and a tetrahedron to a triangle, and the shift to the corresponding number of sides of these 2D-shapes. Presumably, the ambiguity of the term 'side', covering the meanings of both 'face' or 'edge'—depending on the particular setting -, is some underlying cause of this. However, both cases involve some further crucial defect. In the first case, the result does not even equal the correct number of edges. With no own understanding of geometry, edges are simply double counted. In the second case, the sides of the 2D shapes are just taken as the number of the original 3D shapes, resulting in utter nonsense such as "4 sides/cube". However, it is quite remarkable that any of these two wrong patterns—as well as the correct one—are repeatedly and consistently applied to both the shapes, the cube and the tetrahedron, within the same answer, restricting the number of statistically generated values to only three.

A totally different answer distribution is generated for question 8: "Every time the witch has 3 apples, she turns them into 1 banana. Every time she has 3 bananas, she turns them into 1 apple. What will she end up with if she starts with 4 apples and 5 bananas?" Technically, the correct answer, 1 banana, is straightforwardly achieved by a stepwise procedure of some subsequent transformations and does not require any type of higher insight. The necessary considerations actually are only simple number operations that can even be carried out by strokes or with the fingers of both hands. Accordingly, the question belongs to the lowest grade level 1 & 2. On answering that question, ChatGPT regularly generates a pattern of a framework with such a stepwise

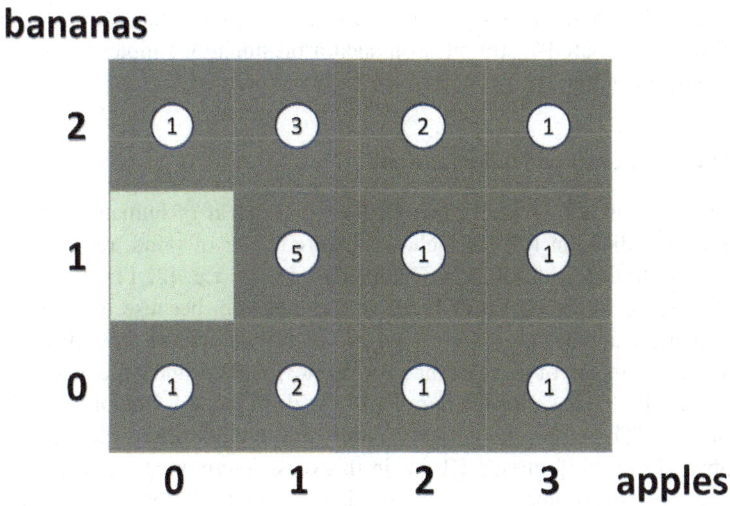

Fig. 5. ChatGPT's answer distribution for 20 independent trials on Kangaroo question 8 displays a huge variety of wrong answers. Every combination of up to 3 apples and up to 2 bananas is generated as an answer at least once, with the actual frequency indicated by the numbers in circles. The correct answer—1 banana only, bright cell -, is the sole exception, not being contained in the set of 20 answers. One further wrong answer "–5 apples" is not depicted in the graphics here.

character. However, within that structure, numbers are not kept up consistently, transformation steps may be falsely executed, or some other inconsistencies and errors occur quite often. Just for illustration, a rather short answer is explicitly stated in the following:

ChatGPT #14:

> "Let's break down the transformations:
>
> 1. Start with 4 apples
>
> – She turns 3 apples into 1 banana. Now she has 1 apple and 1 banana.
>
> – She has 1 apple left.
>
> 2. Start with 5 bananas
>
> – She turns 3 bananas into 1 apple. Now she has 2 bananas and 1 apple.
>
> – She turns 3 bananas into 1 apple. Now she has 1 banana and 2 apples.
>
> – She turns 3 bananas into 1 apple. Now she has 0 bananas and 3 apples.
>
> So, if the witch starts with 4 apples and 5 bananas, she will end up with 3 apples and 0 bananas."

As a result of such a variety of potential options to go wrong, the answer distribution for 20 trials is quite disperse and even displays features similar to a lottery, as depicted in Fig. 5: Every single combination of up to 3 apples and up to 2 bananas is generated in the sample set, except from the correct result, and on top there even exists an answer for the witch to end up with "–5 apples". It must be pointed out, that there is no higher reason for the correct result not to be generated. Thus, for an extended set of trials ChatGPT should certainly come up with "1 banana", too. However, it is quite remarkable, how ChatGPT struggles on such a presumably simple question only involving some clear consecutive steps to be executed correctly.

5.2 Inconsistencies and Hallucinations

Occasionally, ChatGPT produces pieces of text, which if of human origin would be interpreted as indication for a somehow impaired state of mind, ranging from total lack of concentration to insanity, or hallucinations, cf. e.g. [2, 11]. While ChatGPT does not seem to lose the overall focus of the question, because it always provides a final statement about the answer, there are plenty of inconsistent situations where intermediate results are of no relevance for the next parts of the generated answer. For example, question 18 asks for the number of pupils in a class. One of ChatGPT's answers contains: "Therefore, there are 16 pupils in the class. Using Eq. 1, we can find the number of boys and girls: [...] So, in this class, there are 14 girls and 10 boys." One trial of the simple question 33 about pirates climbing a rope goes fatally wrong, because ChatGPT relies on: "Since the captain was the eighth pirate, this means there were three pirates in front of him and three behind him." For the above question about the witch, there are statements such as "The witch also starts with 5 bananas, which cannot be converted into apples since she needs at least 3 bananas to make the transformation." Within two of five trials of question 46, which is about painting a letter each day, ChatGPT dares to claim: "The word KANGAROO has 7 letters.", or "[...] there are 7 letters in the word KANGAROO [...]", respectively. Thus, inconsistencies and even crucial factual mistakes about simple or obvious facts can appear rather locally, within short parts or single sentences of the answer. For most of those shortcomings, also the rather weird "−5 apples" amongst all the 'regular' wrong answers analyzed above, it is normally not possible to really identify underlying reasons or 'lines of thought'. Nevertheless, there are exceptions. As an example, the rather restricted answer distribution of question 48 as discussed above is somehow understandable based on a limited amount of distinct potential errors. Accordingly, some of the weird answers can causally be explained by ChatGPT simply statistically trying out various—also wrong—calculation variants, like a very heterogeneous, very large group of pupils in writing an exam. For example, question 26 requires calculating back from a 2cm high model figure to the size of the real person, and the concrete scale of 1:87 is given. In four of five trials, this works out well. In one of the five trials, however, this scale is applied in the wrong direction, leading to a conclusion not too many human examinees would actually write down in their exam: "Therefore, Ben's brother would be approximately 0.0229 cm tall in real life."

5.3 Luck

An interesting consequence of several independent potential errors within the very same answer is, that ChatGPT might still come up with the correct solution in the final statement, while going wrong even multiple times in the previously generated core text. In a student's exam, such a cancelation of several wrong steps would be considered sheer luck. For example, in the very two cases of question 46 mentioned above, ChatGPT falsely starts from the word 'KANGAROO' to have 7 letters, but nevertheless manages to finally state the correct answer due to an additional mistake in assigning days of the week. In the following, an explicit example is given for question 32: "Marko has 9 pieces of candy and Tomo has 17 pieces of candy. How many pieces of candy does Tomo need to give to Marko so that each boy has the same number of pieces of candy?" This question is assigned to the simplest grade level, and its correct answer is 4, namely half of the difference of Tomos and Markos candies.

ChatGPT #1:

> "In order for Marko and Tomo to have the same number of pieces of candy, they need to divide the total number of candy equally. The total number of candy is 9 + 17 = 26. To equalize the number of candy, Tomo needs to give Marko half of the difference between the total number of candy and Tomo's initial number of candy. Half of the difference is (26—17) / 2 = 4.5. Since we can't have half a piece of candy, Tomo needs to give Marko 4 pieces of candy. Therefore, Tomo needs to give 4 pieces of candy to Marko."

So, after a correct and sensible start ChatGPT follows the trail of a wrong difference, needs to involve a rounding procedure, and is just lucky enough to arrive at the correct result eventually.

6 Resume

As a conclusion, ChatGPT's performance on answering Kangaroo test questions is at best mediocre, if there is no need for real, reliable repeatability. However, if this is taken as a strict requirement as in the setting of this study, the performance is much worse. This indicates ChatGPT for not—yet—being qualified as a stand-alone, all-in-one TRIZ tool in practical application. While the pure statistical results of this study seem consistent with other findings in a comparable setting, cf. [11], they are quite in contrast to e.g. ChatGPT being evaluated as "expert" in the Jeopardy! challenge [12]. Presumably because in the latter, the focus is more—and more simply—about finding and relating facts only. Evaluating answers in more detail reveals, that ChatGPT can handle some rather simple Kangaroo questions quite well—in the sense, that the whole answer text is reasonable, consistent, and correct—in the best case even multiple times in a row. However, this is not the standard. ChatGPT's answer distributions can vary from homogeneous and rather narrow sets to highly disperse varieties of values resembling randomness—even for questions of the lowest grade level. Several types of underlying shortcomings are observed. Apart from ChatGPT losing track of previously generated content, the most crucial ones for the desired kind of TRIZ stand-alone application are certainly pieces of non-factual, non-sensical content, which seem to occur out of nowhere. The step to ChatGPT 4.0 promises a better overall performance, but no general solution to the decisive lack in full reproducibility [11]. An open question is, if and to which extent the performance could be improved involving prompt engineering and further error mitigation strategies, as also reported for other application areas, cf. e.g. [2, 13].

Remark: The full set of 50 test questions—including correct answer, grading, and publication year—is available from the corresponding author upon request.

References

1. Mariani, M.M., Machado, I., Magrelli, V., Dwivedi, Y.K.: Artificial intelligence in innovation research: a systematic review, conceptual framework, and future research directions. Technovation **122**, 102623 (2023). https://doi.org/10.1016/j.technovation.2022.102623
2. Nah, F.F.-H., Zheng, R., Cai, J., Siau, K., Chen, L.: Generative AI and ChatGPT: applications, challenges, and AI-human collaboration. J. Inf. Technol. Case Appl. Res. **25**(3), 277–304 (2023). https://doi.org/10.1080/15228053.2023.2233814
3. Cavallucci, D., Livotov, P., Brad, S. (eds.): Towards AI-Aided Invention and Innovation. Springer Nature Switzerland (2023). https://doi.org/10.1007/978-3-031-42532-5
4. OpenAI. https://openai.com/chatgpt
5. Dolinar, G.: Twenty years of mathematical Kangaroo. EMS Newsletter **85**, 27–28 (2012)
6. Akveld, M., Cáceres, L.: AKSF & math Kangaroo: the world's largest international mathematics competition. Notices Amer. Math. Soc. **69**(11), 1956–1960 (2022)
7. Association Kangourou sans Frontières. https://www.aksf.org
8. Mathematikwettbewerb Känguru e.V. https://www.mathe-kaenguru.de
9. Internal regulations. Association Kangourou Sans Frontières (2019)
10. Math Kangaroo USA, NFP. https://www.mathkangaroo.org
11. Plevris, V., Papazafeiropoulos, G., Jiménez Rios, A.: Chatbots put to the test in math and logic problems: a comparison and assessment of ChatGPT-3.5, ChatGPT-4, and Google Bard. AI **4**(4), 949–969 (2023). https://doi.org/10.3390/ai4040048
12. O'Leary, D.E.: An analysis of Watson vs. BARD vs. ChatGPT: the Jeopardy! Challenge. AI Magazine **44**(3), 282–295 (2023). https://doi.org/10.1002/aaai.12118
13. Biswas, S.: Evaluating errors and improving performance of ChatGPT. Int. J. Clin. Med. Edu. Res. **2**(6), 182–188 (2023)

Developing Function Analysis into the Future

Robert Adunka(✉)

TRIZ Consulting Group GmbH, Sulzbach-Rosenberg, Deutschland
adunka@triz-consulting.de

Abstract. This article highlights function analysis as a central tool within the TRIZ methodology and shows how different approaches can be developed for an adaptable function analysis. It deals with the meta-level of function analysis, domain-specific approaches and the detailed consideration of functions. The article argues in favor of using function analysis in a flexible and adaptable way so that it can meet the specific requirements of different domains.

Keywords: TRIZ · Function Analysis

1 Introduction

Function analysis is one of the basic tools in the TRIZ toolbox [1–3, 5, 6]. Many publications use function models in the analysis phase, which underlines the popularity of this tool. At the same time, however, there is more controversy surrounding this tool than any other. This is due to the different views of the various practitioners of the method. Some find the rules too strict, others too soft. The model is judged to be too vague or too restrictive. This controversy has not yet produced a clear, generally valid function model, which is why an integrative rather than an exclusive approach is advocated here. The postulate for this publication is that the correct function model is the one that will help the user in their consideration of the system and provide insights that can be used in further analyses.

This article deals with finding an adapted approach for each different type of function analysis to avoid discussing the one right type of function analysis. To do this, the article moves to the meta-level to look at what is done in function analysis from there. The concept of domain-specific function analysis is then introduced. This makes it clear that the discussions among the practitioners, which do not lead to the desired result, can be eliminated. Finally, the structure of the functions is detailed and the requirement to change a parameter is examined anew.

O. Mayer (Hrsg.): TRIZ-Anwendertag 2024,
https://doi.org/10.1007/978-3-662-70698-5_9

2 The Meta-Level of Functional Analysis

In TRIZ training courses, the function analysis is presented together with an evaluation of the functions or components. The procedure, known as "function analysis for products", actually consists of two separate steps:

In the first part (modeling), an attempt is made to fit reality into a model with the steps of function analysis: component analysis, (interaction analysis – only for the product) and function modeling. This sometimes works better and sometimes worse, usually depending on the system being modeled.

In a second part (analysis of the model), an analysis view is placed over the model in order to derive innovative questions to be solved, which can then be fed into solution tools. This second part is often done together with the previous step in function modeling, so that the separation is not clearly recognizable.

The best-known aspect of analysis is the incremental improvement of a function model, in which functions are considered and assessed according to usefulness and harmfulness, or the value analysis, in which components are examined and assigned a value (usually functions/costs).

If function analysis is reconsidered and these two steps are treated separately, there are excellent opportunities to optimize both parts and adapt them to external conditions. The result is a modular function analysis made up of building blocks. Adapted to the situation, the building blocks that can make the greatest contribution to the task to be solved are selected.

In this way, different rules can be assigned to different domains when building the model, allowing a more consistent function model to be created within the domain.

In the area of analyzing the model, it is possible to superimpose sustainability, resilience, design or marketing analyses on a function model and derive questions from them. Proof of the feasibility of a sustainability analysis was provided in [13] (Fig. 1).

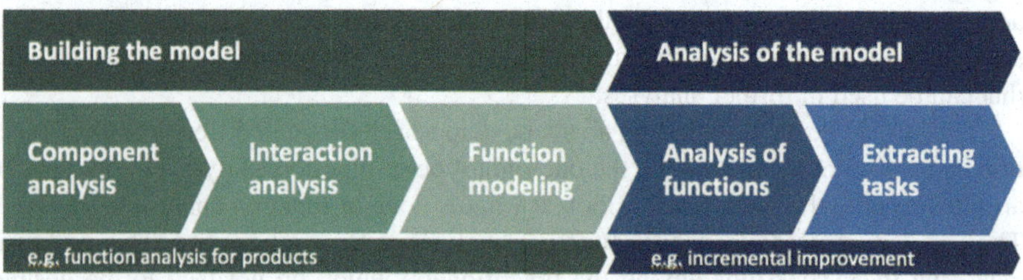

Fig 1: The two parts of the functional analysis

3 Domain-Specific Functional Analysis

The rules for modeling go back to a mechanical view of the world. Some TRIZ masters defend this view, while others try to further develop function analysis for the requirements of the software industry or management. Generally valid rules are then created for these function analyses and their type of function analysis is presented as the only correct one.

In order to be right for all authors of the various function analyses, only the rules for modeling need to be adapted to the domain in which the model is to be formed: In domain-specific modeling, slightly different rules apply to a mechanical system than to electrical, electro-mechanical or information-heavy systems. Software or mechatronics may also have their own requirements for the model. Measurement systems have always caused problems for TRIZ experts and there is a very poor database for function models for chemical systems (Fig. 2).

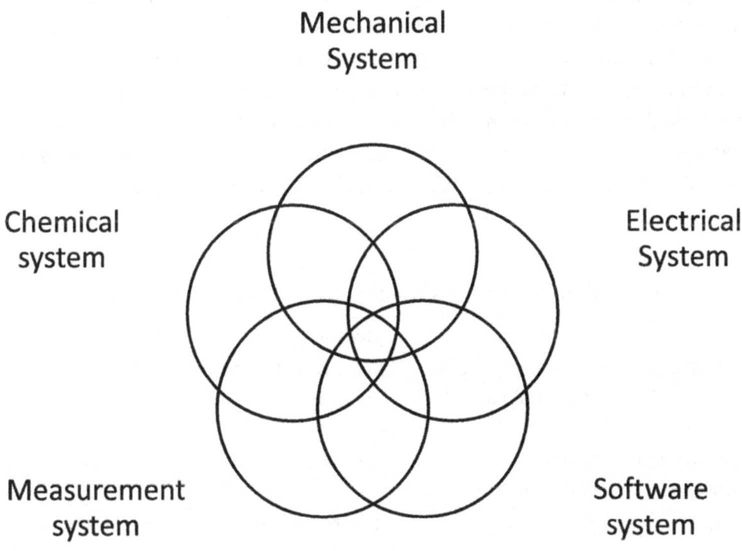

Fig. 2: Domain-specific functional analysis

In order to obtain models that can be used for further analysis and that show the problems of the system in a simple way, the function models and the procedure for obtaining a function model should be adapted to the different domains. This enables a differentiated approach and allows a more flexible response to the customer's requirements.

What rules can be established for the different domains that allow a more consistent and objective modeling of the respective system?

In the rules used in training courses for function analysis, it can be observed that rules are made for the permitted components, it is defined what an interaction is, and it is determined how a function is to be understood. In the case of mechanical systems, for example, this means that components should only have mass, be field-shaped or a combination of both, that interaction exists when components touch each other and that a parameter is changed or maintained due to the function of the object of the function.

In electrical and electromechanical systems, field-shaped components are more frequently included. At PCB level, it is possible to create a function group from a piece of PCB and the components attached to it as an equivalent of an assembly at the mechanical level [4]. Interaction takes place here on a logical level and not as contact, because otherwise all components would touch each other.

Software as a component was not permitted for a long time. However, there are now several publications that aim precisely at integrating software as a component [7–9].

In information-heavy systems, the consideration of the component that executes the function often works the other way around, because one wants to assign the value to the component that is the active part in the functional relationship [11].

Each of these systems has its own characteristics that should be considered when modeling the system. This could be summarized in rules for the different domains. This means that, in absolute terms, a model is no longer better or worse per se than another model but is only better or worse in relation to the use case. Accordingly, the rules to be used when modeling the system should be determined for each domain. Which components are allowed, which are not, what does an interaction mean and when am I allowed to describe a function? Table 1 provides a brief suggestion of what such a set of rules might look like.

Table 1: Characteristics of the models by domain

	Mechanical	Electrical	Software	...
Recommended components	Mass-based, (in exceptional cases: field-based)	Mass-based, field-based, (in exceptional cases: software)	Software (code)	...
Not recommended components	Software, Parameters	Parameters	Parameters	...
Interactions	Touch	Touch or logical interaction	Logical interaction	...
Functions	Changes or maintains "parameters"	Changes or maintains "parameters"	Changes or maintains "parameters"	...

4 Function Viewed in Detail

The requirement for a function was that "A parameter of the object must be changed or maintained". There are three conditions for the existence of a function [14]:

1. function carrier and object of the function must be allowed components (domain-specific restrictions from the component analysis must be observed).
2. the function carrier and the object of the function must interact with each other (domain-specific restrictions from the interaction analysis must be taken into account).
3. at least one parameter of the object of the function must be changed or maintained due to the function of the function carrier.

But what is referred to here as a parameter? Is it a feature, an attribute, a characteristic or a property?

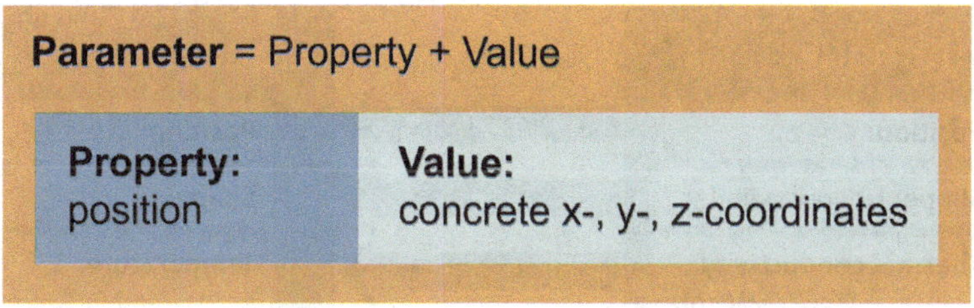

Fig 3: Parameter is property and value.

What was previously referred to as a parameter in the function analysis should be understood as a property or characteristic (see Fig. 3), which can assume a certain value. In the case under consideration, the value of characteristics cannot be changed, and properties can be changed. The properties are then particularly important for the "parameters" in a function analysis, as these can be changed by the function:

For example, a car could have the characteristic "color" with the value "red" and a property "speed" with the value "50 km/h". When creating a functional analysis in which the car is analyzed in its driving state, the speed can be influenced and changed, but the color cannot.

In another analysis, e.g. of the manufacturing process of a car, the characteristic can become a property because the color can be changed here. The combination of property (e.g. color) and value (e.g. blue) thus results in the "parameter". This changes the value of a property based on a function. This is often referred to in simplified terms as changing the parameter.

So, when we talk about a parameter in function analysis, we mean the combination of a property that has assumed a certain value, which can be influenced as part of the function analysis.

This consideration is not new. In OTSM-TRIZ, this model is referred to as the ENV model. The abbreviation ENV stands for "Element – Name – Value". The "element" classically denotes the component, the "name" the property and the "value" the value. In the fractal OTSM ENV model, properties are assigned sub-properties if required. For example, the property "taste" of the component "tomato" can be subdivided into the property "sweetness" and "saltiness" [10].

A function carrier can have several functions on a function object. These different functions change or maintain different properties of the object. A vacuum jug is taken as an example here. Its function is to "hold liquid". If this "holding" is broken down in more detail, it can be seen that there are several "holding" functions. One "hold" function aims to determine the position of the liquid. Another "hold" function maintains the shape of the liquid and a third "hold" function relates to the temperature of the liquid that is to be kept stable (see Fig. 4).

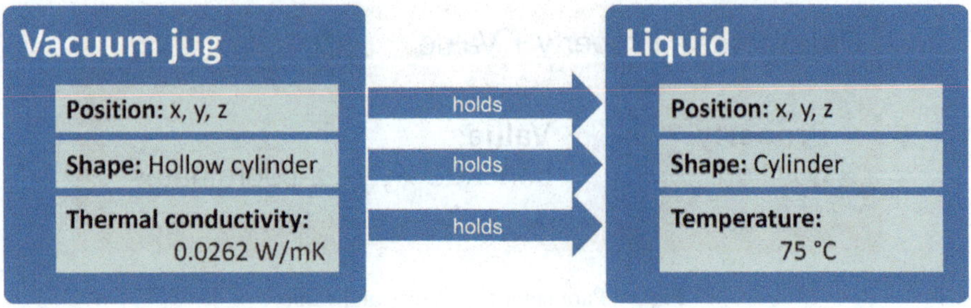

Fig 4: Function “holds” at vacuum jug

If several property values are changed or maintained, the function must be defined for each individual significant change. For example, in the case of the vacuum jug, the second “hold” function could well be omitted because the cylindrical shape of the liquid inside the vacuum jug does not change much. However, it may be important that the liquid is held at all, and that the temperature is kept stable. Modeling is always a simplification of reality. During the modeling process, the creators determine how complex, extensive and realistic the model will be.

This example also shows that there are always properties and characteristics of the function carrier that have an effect on the properties and characteristics of the object of the function. The function, which in this sense can be seen as a transfer function, changes the properties and transfers them to the properties of the object.

If we now assume a transfer function that receives the property values of the function carrier as input and thus changes the values of the properties of the object of the function, we can delve even deeper into physics and model this. If a body is surrounded by air and the body heats the air, then the “heating” can be seen as the transfer function for convection and radiation. It requires certain input values in order to be able to calculate the extent to which the air heats up. If this exact modeling is desired, then a function model could be modeled much more accurately on a physical basis. The function models can then be modeled with openModelica [12] and system physics.

A model of the heat transfer between two solid bodies that are in contact looks like this in openModelica (Fig. 5):

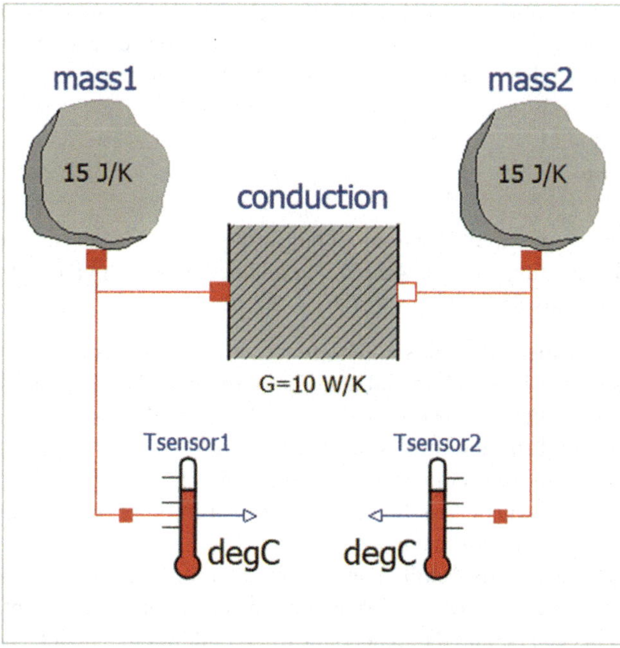

Fig 5: Heat transfer between two solids

The plot over time resulting from the initial parameters entered gives the following picture (Fig. 6):

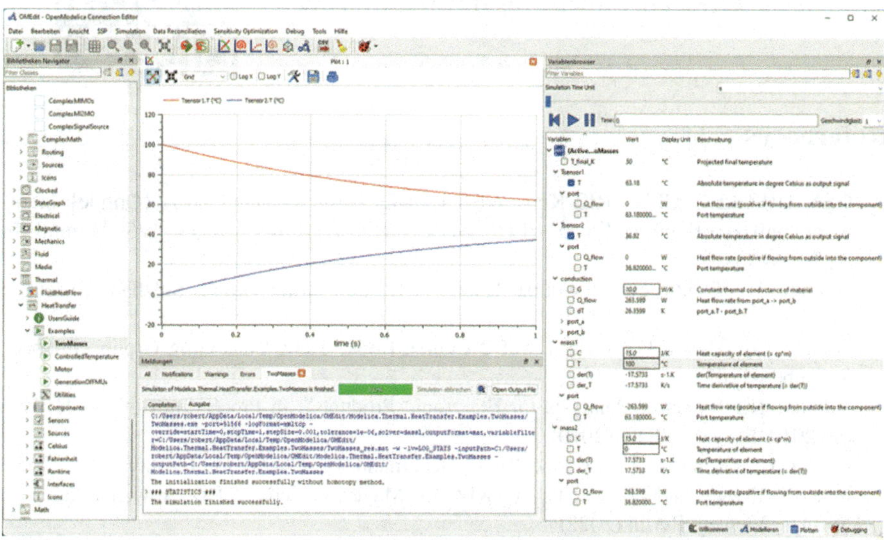

Fig 6: Plot of the heat transfer

From this observation, it quickly becomes clear that the functional modeling sees a snapshot of the system, which is a period of time under consideration. Otherwise, the model could not determine that one body heats the other.

5 Summary

In this publication, possible directions in the further development of function analysis in the field of modeling were presented. An adapted approach can be found for each type of function analysis.

1. **meta-level of function analysis:** The function analysis procedure consists of two separate steps: model building and analysis of the model. By considering these two steps separately, they can be optimized and adapted to the respective requirements.
2. **domain-specific function analysis:** The rules for modeling come from a mechanical worldview, but each domain has its own requirements. The rules for modeling must therefore be adapted to the specific domain in order to maintain models that meet the respective requirements.
3. **function considered in detail:** Parameters are properties and characteristics that can take on a certain value. Different functions can change different properties of an object. This can be taken into account when building the model.

Video of Presentation

The presentation to this paper can be viewed here: https://youtu.be/I7doz7me_Ew?si=dg8EfHizxMDIYtXk

References

1. GEN TRIZ LLC, GEN TRIZ Knowledge Transfer Basic Module (2019). [Online]. https://www.matriz-official.net/images/PDF/GEN_TRIZ_Training_-_Basic_WS_Manual_July_2019pdf.pdf
2. Adunka, R.: TRIZ Level 1 Course Book. TRIZ Consulting Group, Sulzbach-Rosenberg (2023)
3. Adunka, R., Mayer, O.: TRIZ Level 2 Course Book. TRIZ Consulting Group, Sulzbach-Rosenberg (2023)
4. Adunka, R.: "Function Analysis for Electronic Products", in Proceedings of the TRIZFuture2010. Bergamo, Italy (2010)
5. Nähler, H., Gronauer, B., Bertoncelli, T., Beckmann, H., Chrząszcz, J., Mayer, O.: Modelling of software and IT systems in TRIZ. In: Mayer, O. (eds.) TRIZ-Anwendertag 2022. Springer Vieweg, Berlin (2022)

6. Chrząszcz, J., Bertoncelli, T., Gronauer, B., Oliver Mayer, Nähler, H.: Modeling IT systems in TRIZ. In: Nowak, R., Chrząszcz, J., Brad, S. (eds.) Systematic Innovation Partnerships with Artificial Intelligence and Information Technology. TFC 2022. IFIP Advances in Information and Communication Technology, vol. 655. Springer, Cham (2022)
7. Nähler, H., Gronauer, B., Bertoncelli, T., Beckmann, H., Chrząszcz, J., Mayer, O.: Modeling software in TRIZ. In: Nowak, R., Chrząszcz, J., Brad, S. (eds.) Systematic Innovation Partnerships with Artificial Intelligence and Information Technology. TFC 2022. IFIP Advances in Information and Communication Technology, vol. 655. Springer, Cham (2022)
8. Khomenko, N.: See Video https://youtu.be/oGZaqytGXKs?si=E-1K-Q1TyclIKhG6 at 58:00, slides 74 ff.
9. Souchkov, V.: „Extended Function Analysis", gehalten auf der Expertentag 2019. Sulzbach-Rosenberg (2019)
10. https://build.openmodelica.org/Documentation/Modelica.Thermal.HeatTransfer.Examples.TwoMasses.html
11. Mayer, O.: Nachhaltigkeitsmodellierung mit TRIZ. In: Mayer, O. (eds.) TRIZ-Anwendertag 2022. Springer Vieweg, Berlin (2022)
12. Guedes, L.C.: „TRIZ And Software Evolution", in Proceedings of the MATRIZ TRIZfest 2018 International Conference. Lisbon, Portugal (2018)
13. Litvin, S., Prokofiev, A.: „TRIZ Approach To Internet Of Things", in Proceedings of the MATRIZ TRIZfest 2018 International Conference. Lisbon, Portugal (2018)
14. Lyubomirskiy, A., Litvin, S.: GEN TRIZ Basic Module. GEN TRIZ, LLC. (2024)

Idealität vor dem Hintergrund Human-Centered-Design und Nachhaltigkeit

Uwe Schaumann[1] und Oliver Gerundt[2](✉)

[1] E.G.O. GmbH, Oberderdingen, Deutschland
uwe.schaumann@egoproducts.com
[2] Robert Bosch GmbH, Gerlingen-Schillerhöhe, Deutschland
oliver.gerundt@de.bosch.com

Zusammenfassung. Während der Produktentwicklung wird die Wirtschaftlichkeit, Nutzenerfüllung sowie Nachhaltigkeit von Lösungen bewertet. Bei der Wirtschaftlichkeit treten dabei häufig Widersprüche auf. Um eine quantitative Vergleichbarkeit und Transparenz herstellen zu können, wird eine Funktionsbewertung vorgestellt, die in eine „Wertanalytische Betrachtung“ einfließt. Mithilfe der Idealität nach TRIZ können Funktionen quantifiziert werden, wodurch Lösungen unter Berücksichtigung von Nachhaltigkeit und Nutzererwartungen unter wirtschaftlichen Gesichtspunkten verglichen werden können.

Schlüsselwörter: TRIZ · Idealität · Nachhaltigkeit · Human-Centered Design

1 Ausgangssituation

Nachhaltige Lösungen spielen in der Produktentwicklung eine zentrale Rolle. Das gilt für alles drei Aspekte der Nachhaltigkeit: soziale, wirtschaftliche und ökologische Nachhaltigkeit. Durch diese unterschiedlichen Schwerpunkte werden meist widersprüchliche Anforderungen an ein Produkt gestellt, was zu gegenläufigen Vor- und Nachteilen des Produkts führt. Insbesondere die wirtschaftliche Nachhaltigkeit wird häufig durch Lösungen im Bereich Ökologie und Soziales geschwächt [8]. Im Allgemeinen ist die Zahlungsbereitschaft der Kunden für nachhaltige Produkte gering, weshalb die Kundensegmente gezielt betrachtet werden müssen [5].

Ein weiterer Schwerpunkt in der Entwicklung von Produkten ist die Kundenzentrierung, das sogenannte Human-Centered Design (nachfolgend HCD). Hierbei wird nicht nur die Benutzung des Produkts betrachtet, sondern der gesamte Zyklus von der Informationsbeschaffung, dem Kauf, über die Benutzung, Reparatur bis zur Entsorgung, die sogenannte „Customer Journey“.

Und hier kommt nun die Verbindung der Themen Human-Centered Design und Nachhaltigkeit ins Spiel. Beispielsweise bekommt die Reparierbarkeit eine wachsende Bedeutung, da sie die Lebensdauer des Produkts verlängert und somit Rohstoffe und Herstellungsenergie einspart. Da die Ermöglichung von Reparierbarkeit meist mit

O. Mayer (Hrsg.): TRIZ-Anwendertag 2024,
https://doi.org/10.1007/978-3-662-70698-5_10

höheren Kosten einhergeht, für die Kunden eine beschränkte Zahlungsbereitschaft zeigen, kommt der Interpretation der Idealität eine ganz neue Bedeutung zu.

1.1 Nachhaltigkeit

Nachhaltigkeit spielt seit vielen Jahren eine zunehmende Rolle im Design für Produkte, Produkt-Service-Systeme, räumlich-soziale und sozio-technische Systeme [3]. Dabei ist eine schrittweise Erweiterung von einem technischen und produktorientierten Fokus zu Veränderungen auf Systemebene, bei denen Nachhaltigkeit als sozio-technische Herausforderung verstanden wird, sichtbar. Damit steigt die Komplexität, was sich auch in einer Vielzahl von Methoden für die Analyse bspw. Lebenszyklusanalyse (LCA) und Lösungsfindung für die Produktentwicklung zeigt [4].

Ein wichtiger Aspekt für die Produktentwicklung ist die Bewertung von technischen Funktionen. Ein Weg hierfür bietet die TRIZ-Funktionsanalyse (FA) in Kombination mit einer PESTEL-Analyse [12], die für jede Entwicklungsaufgabe zu Beginn durchgeführt werden kann [9]. Damit können Nachhaltigkeitsziele auf das betrachtete System so heruntergebrochen werden, dass Entwicklungsrichtungen, Kriterien bzw. Widersprüche deutlich werden.

Mit der PESTEL-Analyse können die folgenden Elemente betrachtet werden [2]:

- Soziokulturelle Faktoren wie Werte, Lebensstil
- Technologische Faktoren wie Forschung, Produktlebenszyklus
- Ökonomische Faktoren wie Verfügbarkeit von Ressourcen
- Ökologische Faktoren wie geografische Faktoren, Umweltschutz
- Politische Faktoren wie Gesetzgebung, Sicherheitsvorgaben

Soll zusätzlich der Nutzen für verschiedene Stakeholder und die damit verbundenen Geschäftsmodelle mitberücksichtigt werden, kann die PESTEL-Analyse mit einer Value Mapping ergänzt werden [12].

Im Ergebnis liegen Aufgabenstellungen vor, die mit TRIZ in Richtung einer idealen Lösung weiterbearbeitet werden können. Für das Fallbeispiel wurden die Nachhaltigkeitsziele Ökologie, Wirtschaftlichkeit und Soziales um die Aspekte Technologie und Gesetzgebung erweitert. Für das Fallbeispiel wurde eine Vereinfachung vorgenommen (Abb. 1 und 2).

Mit der Auswahl des durch Gesetze geforderten Ziels der Verbesserung der Reparaturfähigkeit kann ein Rahmen für die Ableitung von technischen Fragen mithilfe der Funktionsanalyse (FA) erfolgen, was in Abschn. 2.1. für die Funktion „Halten" beschrieben wird. Sowohl mit der PESTEL-Analyse als auch mit der Funktionsanalyse sind konkrete Kriterien für die Bewertung definierbar [13, 14].

1.2 Human-Centered Design

Human-Centered Design (HCD) zielt darauf ab, Produkte optimal auf die Menschen bzw. Nutzer abgestimmt zu entwickeln. Die Anforderungen und Bedürfnisse stehen dabei im Mittelpunkt. Dazu zählen nicht nur Funktionen, sondern auch Ergonomie und Benutzerfreundlichkeit sowie eine intuitive Bedienung bzw. Nutzung. HCD

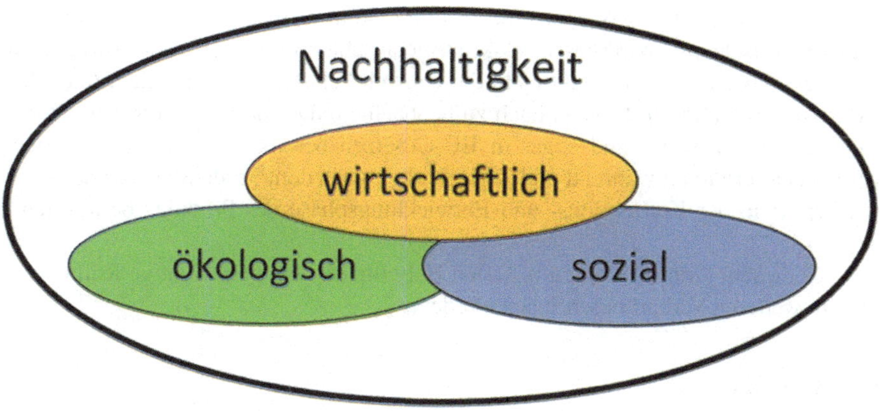

Abb. 1: Aspekte der Nachhaltigkeit

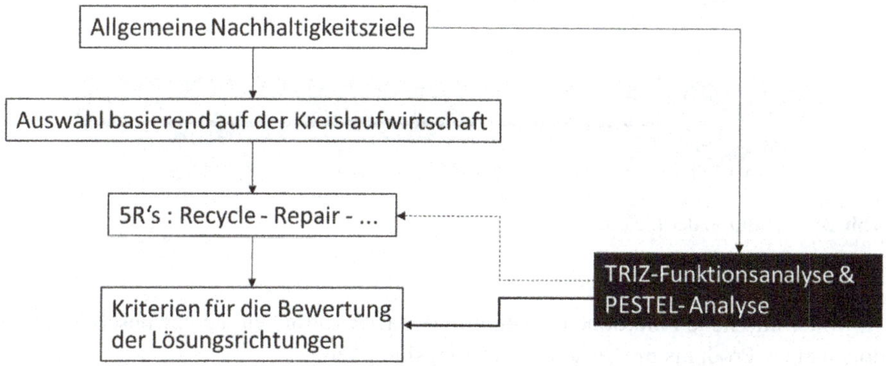

Abb. 2: Kriterien Auswahl mit PESTEL- und Funktions-Analyse

könnte somit auch als auf den Menschen abgestimmter, ganzheitlicher Entwicklungsansatz bezeichnet werden. Bei einigen der Methoden wird zwischen Kunden, Nutzern und Menschen im Allgemeinen unterschieden. Da die Auswirkungen der Details stark vom Produkt abhängen, soll im weiteren Verlauf der Einfachheit und Lesbarkeit halber von Nutzern geschrieben werden.

In der praktischen Umsetzung bedeutet das einige Änderungen und zusätzliche Schritte im Entwicklungsprozess. Ohne auf diese im Detail einzugehen, ist ein Werkzeug die sogenannte Customer-Journey. Hier wird aus der Perspektive der Nutzer der gesamte Prozess beschrieben, in dem die Nutzer mit dem Produkt zu tun haben. Das beginnt beim Informieren über das Produkt und die Alternativen und beim Kauf. Beim Kauf spielen „weiche" Faktoren eine wichtige Rolle. Die Nutzer wollen ein gutes Gefühl haben, das richtige Produkt gekauft zu haben. Auch das Thema Nachhaltigkeit trägt zu diesem Gefühl bei. Während der Nutzung stehen Eigenschaften wie Ergonomie, intuitive Bedienung und die Funktionen allgemein im Vordergrund.

Im Falle eines Defekts oder einer Störung stellt sich die Frage nach Austausch oder Instandsetzung. Wird das Produkt ausgetauscht, so möchten viele Nutzer sicher sein, dass das entsorgte Produkt nicht die Umwelt belastet, sondern einem Recycling-Kreislauf zugeführt wird. Es ist auch zu beobachten, dass das Thema Reparierbarkeit an Bedeutung gewinnt und sogar in EU-Gesetzen bereits teilweise vorgeschrieben wird. Reparierbarkeit während und Zerlegbarkeit nach dem Produktlebenszyklus müssen bereits in der Konzeptions- und Entwicklungsphase des Produkts berücksichtigt werden.

Aus diesem Grund kommen u. a. den Haltefunktionen eine wichtige Rolle zu, die hier im weiteren Verlauf beschrieben werden.

1.3 Idealität

Der Begriff der Idealität kommt aus dem TRIZ-Werkzeugkasten. Mit Idealität wird die dimensionslose Qualitätsgüte eines Systems ermittelt. Sie setzt sich zusammen aus dem Quotienten des Nutzens und der nützlichen Funktionen durch die Summe der Kosten und negativen Effekte (Abb. 3).

$$\text{Idealität} = \frac{\Sigma\ \text{nützliche Funktionen} + \Sigma\ \text{positive Effekte}}{\Sigma\ \text{Kosten} + \Sigma\ \text{negative Effekte}}$$

Abb. 3: Definition der Idealität

Unter nützliche Funktionen versteht man im Wesentlichen die gewünschten Funktionen eines Produkts mit all seiner Leistungsfähigkeit.

Die positiven Effekte fassen die eher nicht-materiellen Merkmale eines Produkts zusammen, wie Qualitätsanmutung, Image und Nutzerbegeisterung, aber auch positive Einflüsse, Schaffung von Arbeitsplätzen, positive Wirkungen auf die Umwelt, usw.

Die Kosten fassen die Kosten für Material und Halbzeuge oder Bauteile von Zulieferern und die Kosten für die Wertschöpfung und Gemeinkosten zusammen.

Unter den negativen Effekten versteht man im Allgemeinen Dinge wie negative Auswirkungen auf die Umwelt, Ausbeutung von Menschen, Emissionen, usw.

Das Streben nach höherer Idealität ist der Antrieb für jegliche Evolution technischer Systeme. Abhängig von der Reife eines Produkts treten andere Kriterien aus der Formel in den Vordergrund. Während zu Beginn des Produktlebens die Funktion und Performance den Schwerpunkt in der Entwicklung bilden, sind es gegen Ende die Kosten.

Das Modell der Idealität kann nicht nur als eine Art Gradmesser für die Güte dienen, sondern es können daraus auch Entwicklungsrichtungen und Kriterien für die Auswahl technischer Lösungen abgeleitet werden.

2 Vorgehensweise

Die Erzeugung eines Funktionsmodells läuft gleich wie beim klassischen TRIZ ab und wird an dieser Stelle nicht weiter erläutert [15]. Zusätzlich kann eine PESTEL Umfeldanalyse durchgeführt werden, um die Einflussfaktoren aus Politik, Wirtschaft, Sozialem, Technologie, Gesetzgebung und Umwelteinflüssen zu analysieren. Anders als beim klassischen TRIZ wird die Wertigkeit der Funktionen individuell vergeben.

Nachhaltigkeit und Human-Centered Design (HCD) sind Elemente, die im Produktentwicklungsprozess berücksichtigt werden sollten. Ein Produkt besteht meist aus vielen einzelnen Bauteilen, die u. a. mechanisch verbunden werden müssen. Um ein Produkt reparieren zu können und um es nach seiner Nutzungsdauer in Einzelteile zerlegen zu können, sollten das Produkt so einfach wie möglich zerlegbar sein. An dieser Stelle kommen die Haltefunktionen ins Spiel und die Anforderungen daran. Aus Kostengründen ist häufig eine nicht-zerlegbare Verbindung günstiger. Hier seien Löten für elektrische und Kleben für mechanische Verbindungen genannt. Betrachtet man die Idealität ohne Nachhaltigkeit und HCD, hat man einen guten Stand erreicht, da Kleben und Löten kostengünstige Verfahren sind und die nützlichen Funktionen bestens erfüllen. Bezieht man den Wunsch der Nutzer jedoch mit ein, dass sie ihr Produkt lange nutzen und mit dessen Entsorgung die Umwelt nicht belasten wollen, so ist die Idealität des Produkts durch größere negative Effekte im Nenner schlechter geworden.

Das heißt, dass sich die Kriterien, warum ein bestimmtes Verbindungsverfahren ausgewählt wird, verändern.

2.1 Bewertung der Funktionen anhand eines Beispiels

Stand der Technik bei der Kontaktierung von Akkus in Kleingeräten ist das Anlöten der Leitungen an sogenannte Lötfahnen. Das ist eine kostengünstige Möglichkeit. Die Verbindung ist jedoch nur wieder durch Löten lösbar, was in einem Reparaturfall bzw. beim Trennen der Bauteile beim Recycling am Ende des Produktlebenszyklus für größeren Aufwand sorgt und vertiefte Kenntnisse sowie eine umfangreichere Ausstattung erfordert. Das folgende Beispiel zeigt diese Technik anhand eines kleinen Elektrowerkzeugs (Abb. 4).

Als Basis für die vereinfachte Funktionsanalyse dient folgende Prinzipskizze (Abb. 5).

Das abgeleitete Funktionsmodell ist im nächsten Bild zu sehen. Die Ansicht ist stark vereinfacht dargestellt, um die wesentlichen Funktionen hervorzuheben. Zusätzlich wurden die Funktionen nach klassischer TRIZ ‚Wertanalytischer Betrachtung‘ bewertet (Abb. 6).

Einfachheitshalber ist die Funktion „Elektronik versorgt Antrieb“ als Basisfunktion dargestellt. Auffällig ist die unterschiedliche Bewertung der Haltefunktionen, die von A1 bis A3 reicht. In der Darstellung ist die Funktion „Gehäuse hält Elektronik“ um zwei Stufen höherwertiger als die Funktion „Gehäuse hält Akku“. Im Fall einer Reparatur ist das jedoch nicht logisch, denn dann müsste allen Haltefunktionen, die für den Austausch des Akkus notwendig sind, eine höhere Wichtigkeit und damit

Abb. 4: Gelötete Anbindung eines Kleingeräteakkus

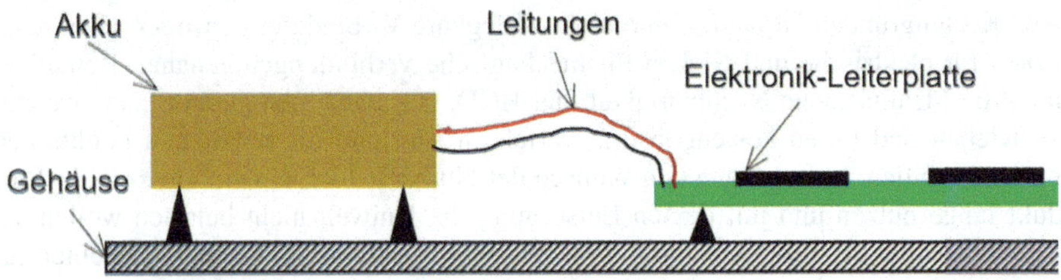

Abb. 5: Prinzipdarstellung

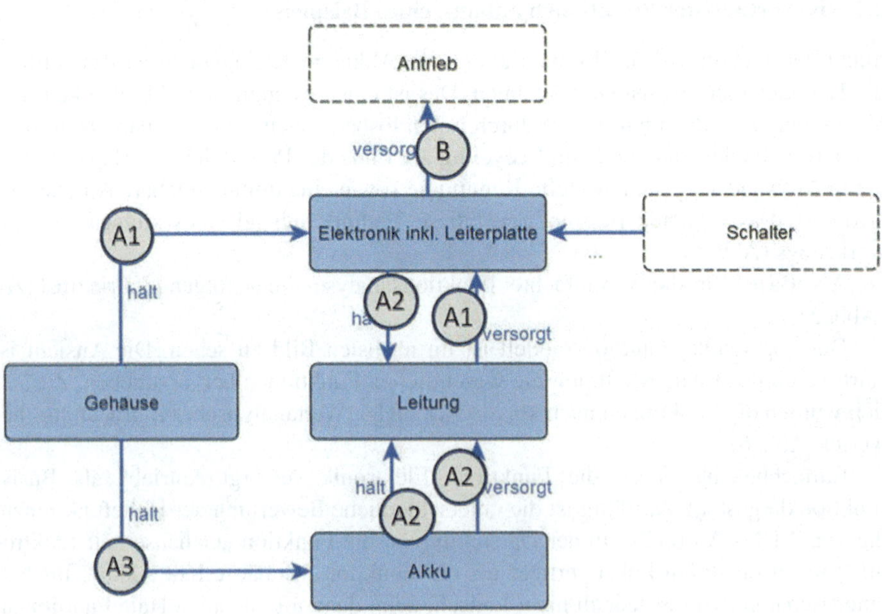

Abb. 6: Funktionsmodell (vereinfachte Darstellung) mit Bewertung

eine höhere Wertigkeit zugeordnet werden, da die Wahrscheinlichkeit des Ausfalls des Akkus höher als die der Elektronik ist. Damit müssten die Funktionen „Gehäuse hält Akku" und „Akku hält Leitung" zumindest die gleiche Wertigkeit bekommen.

Die Ränge können mit einem konkreten Wert versehen werden, was in folgender Tabelle dargestellt ist (Tab. 1).

Tab. 1: Zuordnung Funktionswerte zu Funktionsrängen

Funktionsrang	Funktionswert
B	5
A1 = Ad	3
A2	2
A3	1

Das Beispiel aus einem anderen Kleingerät einer späteren Generation zeigt im nächsten Bild für die gleiche Anwendung der Akkuanbindung eine andere Konstruktion. Hier ist die Akkuzelle mit Blechzungen versehen, auf die die Leitungen mit Kabelschuhen angesteckt sind (Abb. 7).

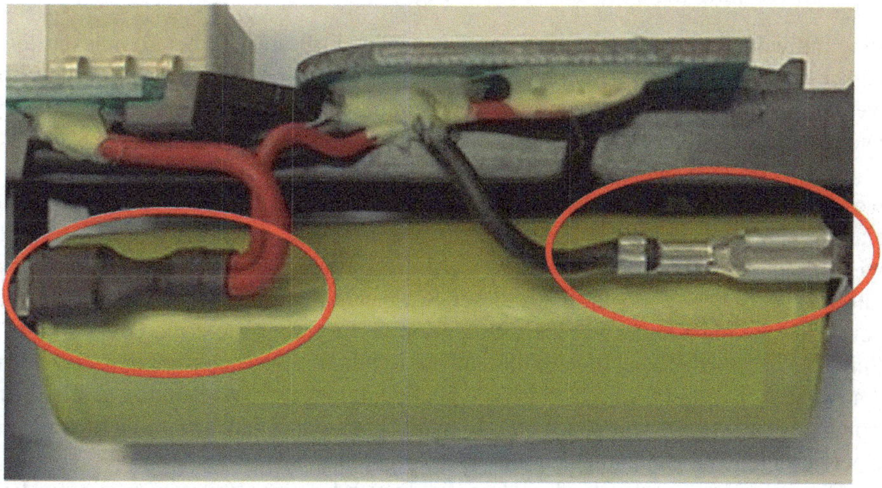

Abb. 7: Beispiel einer gesteckten Anbindung eines Geräteakkus

Die gezeigte Lösung ist teurer und müsste eigentlich nach klassischer Bewertung, ohne Nachhaltigkeitskriterien, verworfen werden.

2.2 Notwendigkeit neuer Bewertungskriterien

Bei allen Vorteilen der klassischen „Wertanalytischen Betrachtung" nach TRIZ, stößt sie bei einigen Beispielen an ihre Grenzen (s. a. [6, 11]). Insbesondere in Fällen, bei denen zusätzliche Kriterien und Anforderungen von außen an das System gestellt werden. Ein „gutes Gefühl" des Nutzers oder gesetzliche Bestimmungen lassen sich nur schwer in einem klassischen Systemmodell abbilden. Aus diesem Grund macht eine individuelle Vergabe anhand von Kriterien Sinn. Die praktische Umsetzung der Bewertung mittels paarweisen Vergleichs ist in [11] detailliert beschrieben und wird hier nicht weiter ausgeführt. Nur so viel, dass das Ranking der Funktionen mit der Methode „Paarweiser Vergleich" gemacht wird. Hierbei wird jede Funktion mit allen anderen verglichen und bewertet, welche der beiden wichtiger ist oder ob sie gleichwertig sind. Hierzu sind Kriterien hilfreich, insbesondere, wenn es sich um ein komplexes Umfeld handelt. In der folgenden Tabelle sind die Kriterien aufgelistet, die die Erfahrungen aus mehreren Praxisanwendungen widerspiegeln. Die erste Tabelle dieser Art, die in [10] beschrieben ist, wurde hier um die Kriterien der Nutzerzentrierung und Nachhaltigkeit ergänzt.

Kriterien für höhere Priorität sind unter „>" geclustert, gleichwertige unter „=" und weniger wichtige unter „<" (Tab. 2).

2.3 Anwendung der Funktionsbewertung

Eine individuelle Bewertung von Funktionen führt insbesondere bei komplexen Systemen zu Schwierigkeiten bei der Durchführung. Die Komplexität kann vereinfacht werden, indem jeweils nur zwei Funktionen miteinander verglichen werden [10, 11]. Die folgende Tabelle zeigt einen solchen Vergleich für das Beispiel (Tab. 3).

Hier zeigt sich durch die Anwendung der Kriterien ein realistischeres Bild. Die wertigste Funktion ist die Basisfunktion mit 12 Punkten. Das ist gleichzeitig die Hauptfunktion, denn dafür werden Elektronik, Akku, usw. verbaut. Am zweitwichtigsten sind die beiden Funktionen, die es ermöglichen, den Akku im Schadensfall auszutauschen (9 Punkte). An dieser Stelle wirkt sich aus, dass die Nachhaltigkeitskriterien und Kundenwünsche hoch priorisiert werden. Die dritte Gruppe (4 & 5 Punkte) sind wieder Funktionen, die auf die Erfüllung der Hauptfunktion einzahlen. Am wenigsten relevant sind die internen Haltefunktionen (1 & 2 Punkte), da die Elektronik im Gegensatz zum Akku eine höhere Lebensdauer hat und Bauteilfehler mit einer geringen Wahrscheinlichkeit betrachtet werden.

Damit konnte gezeigt werden, dass die Anwendung der Kriterien in diesem Fall zu einem realitätsnäheren Bild führen als die klassische Funktionsbewertung, wie sie unter Abschn. 2.1 angewendet wurde. Im weiteren Verlauf würden die Ergebnisse in eine Kostenanalyse münden, wie sie in TRIZ bei der „Wertanalytischen Betrachtung" durchgeführt wird.

2.4 Anwendung der Idealität

Als Hilfsgröße zur Konzeptentscheidung ist die Idealität eine geeignete Größe. Das Problem ist nur, dass sie eine dimensionslose Größe ist und damit eine Entscheidung beim direkten Vergleich der Idealität zweier Konzepte erschwert wird (Abb. 8).

Tab. 2: Bewertungskriterien für Funktionen

	Kriterium	Beschreibung
„>"	Basisfunktion	Basisfunktionen haben im Allgemeinen eine höhere Priorität als alle anderen Funktionen
	Nähe zur Basisfunktion/ Funktions-hierarchie	Funktionen, die direkt dazu dienen, die Hauptfunktion zu erfüllen, werden als wichtiger betrachtet als Funktionen in der Peripherie Z. B. zählen bei einem Akkuschrauber dazu alle Komponenten des Antriebsstrangs, vom Akku, über Motor, Getriebe bis zum Futter
	Funktionsautonomie	Kann die Komponente diese Funktion selbst erfüllen oder braucht sie die Unterstützung einer nachgeordneten Komponente? Z. B. Motor ist wertiger als Getriebe, da die essentielle rotatorische Energie vom Motor bereitgestellt wird
	Nutzer-zentrierung/ Begeisterungs-funktionen	**Begeisterungsfunktionen sind als sehr hochwertig zu bewerten, da dies zur Entwicklung nutzer- zentrierter Produkte unerlässlich ist.** **Mit Begeisterungsfunktion ist gemeint, inwieweit Nutzer in positiver Weise von Komponenten angesprochen werden, beispielsweise der Haptik eines Produkts oder dem Geräusch, das ein Antrieb verursacht** **Für die Bewertung ein komplexes Thema, aber inwieweit trägt die Funktion dazu bei, beim Nutzer ein gutes Gefühl zu erzeugen?** **Fällt ein Merkmal hier negativ auf, in dem Sinne, dass es den Nutzer stört, so ist die Funktion als schädlich (harmful) zu kategorisieren. Genannt sei hier ein unangenehmes Geräusch oder Geruch nach billigem Plastik**
	Erhöhte Anforderungen aufgrund von z. B. Nachhaltigkeits-aspekten	Werden erhöhte Anforderungen an eine Funktion gestellt, kann eine höhere Gewichtung zugewiesen werden Z. B.: **Ist eine Haltefunktion innerhalb eines Systems lösbar, um Reparierbarkeit oder das Zerlegen beim Recycling zu ermöglichen, so sind diese höher zu bewerten**
	Hauptnutzungs-zeit	Funktionen in der Hauptnutzungszeit sind wertiger als Funktionen eines anderen Zeitpunkts. Z. B.: Bei einem Schrauber ist das Schrauben wichtiger als das Laden
	Zusatz-funktionen	Zusatzfunktionen (Ad) werden höher bewertet als einfache Haltefunktionen
	Gleicher Funktionsträger	Haben zwei Funktionen denselben Funktionsträger, dann ist der Beitrag zur Hauptfunktion des Systems entscheidend
„="	Gleiche Kriterien	Wenn das gleiche Kriterium für zwei Funktionen gilt, die miteinander verglichen werden, so werden diese als gleich wichtig bewertet Beispiele: → Haltefunktionen für die Komponenten in einem System **→ Begeisterungsfunktionen für den Nutzer, die diesen in gleicher Weise ansprechen. Ist eine davon weniger stark, dann wird sie auch als weniger wichtig bewertet**
	Über-schneidungen der Kriterien	Wenn zwei Kriterien sich überschneiden und keine einheitliche Entscheidung getroffen werden kann, wird der Vergleichswert 1 vergeben
„<"	Zweitrangige Zusatz-funktionen	Dies gilt auch für Funktionen, die an der Zielkomponente sind

Tab. 3: Paarweiser Vergleich der Funktionen

	Elektronik versorgt Antrieb	Leitung versorgt Elektronik	Akku versorgt Leitung	Gehäuse hält Akku	Gehäuse hält Elektronik	Elektronik hält Leitung	Akku hält Leitung	Summe Funktionalität	Beitrag an die Gesamtfunktion
Elektronik versorgt Antrieb		2	2	2	2	2	2	12	29%
Leitung versorgt Elektronik	0		1	0	2	2	0	5	12%
Akku versorgt Leitung	0	1		0	1	2	0	4	10%
Gehäuse hält Akku	0	2	2		2	2	1	9	21%
Gehäuse hält Elektronik	0	0	1	0		1	0	2	5%
Elektronik hält Leitung	0	0	0	0	1		0	1	2%
Akku hält Leitung	0	2	2	1	2	2		9	21%

$$\text{Idealität} = \frac{\Sigma\ \text{Positives (vorher)} + \text{Reparierbarkeit}}{\Sigma\ \text{Negatives (vorher)} + \text{Kosten Zusatzmaterial} + \text{-prozesse}}$$

Abb. 8: Idealität neu

Für das gezeigte Beispiel der Akkuanbindung ist es so, dass zusätzliche Kosten durch Material und zusätzliche Prozesse entstehen, aber eben auch ein großer Zusatznutzen wie die Reparierbarkeit durch vereinfachte Montageprozesse. Die Kontaktierung des gesteckten Akkus funktioniert komplett ohne Werkzeug. Im Sinne der Nachhaltigkeit ist der Nutzen der Reparierbarkeit höher anzusehen als die zusätzlichen Kosten. Dabei verwundert es, dass auf dem Produkt nicht explizit auf die Reparaturfähigkeit hingewiesen wird. Denn dies würde doch zu einem guten Gefühl bei den Käufern und für einen besseren Reparaturindex sorgen.

Nun stellt sich die Frage nach der praktischen Anwendung der Erkenntnisse. Einerseits wurde gezeigt, dass sich der dimensionslose Quotient der Idealität als Mittel eignet, zusätzliche Kriterien aus dem ‚Human-Centered Design' und der Nachhaltigkeit zu berücksichtigen. Andererseits sind die Quotienten unterschiedlicher Lösungen durch ihre Dimensionslosigkeit nicht direkt miteinander vergleichbar. Um die Vergleichbarkeit zu ermöglichen, wird folgendes Verfahren vorgeschlagen: Die Werte in Zähler und Nenner des Quotienten bleiben gleich, wobei das Merkmal, für das es unterschiedliche Designmerkmale zur Entscheidung gibt, ausgeklammert wird. Dann werden Deltawerte hinzugefügt, um die Konzepte miteinander vergleichen zu können. Tab. 4 zeigt die unterschiedlichen Möglichkeiten, wobei „I" die Idealität bedeutet.

Tab. 4: Deltawerte für die Idealität

$I=\frac{1}{1}$	$I=\frac{-}{1}$	$I=\frac{+}{1}$	$I=\frac{1}{-}$	$I=\frac{1}{+}$
nicht relevant oder neutral	weniger positive Effekte → I sinkt	mehr positive Effekte → I steigt	weniger schädliche Effekte → I steigt	mehr schädliche Effekte → I sinkt

Die Ausdrücke sind wie folgt zu interpretieren:

$I=\frac{1}{1}$ bedeutet, dass das Kriterium nicht relevant oder neutral zu bewerten ist.

$I=\frac{-}{1}$ bedeutet, dass die positiven Effekte weniger stark ausgeprägt sind, wodurch die Idealität sinkt.

$I=\frac{+}{1}$ bedeutet, dass die positiven Effekte stärker ausgeprägt sind, wodurch die Idealität steigt.

$I=\frac{1}{-}$ bedeutet, dass die negativen Effekte weniger stark ausgeprägt sind, wodurch die Idealität steigt.

$I=\frac{1}{+}$ bedeutet, dass die negativen Effekte stärker ausgeprägt sind, wodurch die Idealität sinkt.

Fallen die Ausschläge größer aus, so können „+“ und „-“ mit dem Faktor zwei oder drei versehen werden.

Für die Aspekte, die für die Deltabetrachtung der Idealität in Betracht gezogen werden, wird die PESTEL-Analyse herangezogen. In der PESTEL-Analyse werden die aus der Nachhaltigkeit bekannten Aspekte Wirtschaftlichkeit, Soziales und Ökologie um die Aspekte Technologie und Gesetzgebung ergänzt. Für diese fünf Kategorien wird jeweils für die Entscheidungsdispositionen eine Deltabetrachtung gemacht. Tab. 5 zeigt dies am Beispiel Akkuanbindung.

Tab. 5: Vergleichbarkeit der Idealität

$I=\frac{\text{Nützliches}}{\text{Schädliches}}$	sozial	ökologisch	wirtschaftlich	technologisch	gesetzlich	Σ
Löten	$I=\frac{1}{1}$ nicht relevant	$I=\frac{-}{1}$ Recycling	$I=\frac{1}{2-}$ kostengünstiger	$I=\frac{-}{1}$ bleifreie Lote	$I=\frac{-}{1}$ Zuk. Auflagen	$I=\frac{3-}{2-}=1-$
Stecken	$I=\frac{1}{1}$ nicht relevant	$I=\frac{2+}{1}$ Reparierbarkeit & längere Lebensdauer	$I=\frac{1}{2+}$ höhere Montagekosten	$I=\frac{+}{1}$ robust, einfach	$I=\frac{+}{1}$ erfüllt zukünftige Gesetzgebung	$I=\frac{4+}{2+}=2+$

Die Veränderung der Idealität macht verschiedene Konzepte damit vergleichbar, obwohl die Idealität ein dimensionsloser Quotient ist. Die hier gezeigten fünf Kategorien sind an dieser Stelle sehr grob gewählt und können je nach Thema weiter verfeinert, ergänzt oder sogar weggelassen werden. Das Thema „Human-Centered Design“ kann verschiedenen der gezeigten Cluster zugeordnet werden. Die Funktionen, die die Anwendung des Produkts für den Nutzer intuitiver und einfacher machen, können unter dem Cluster „technologisch“ einsortiert werden. Der Kundenwunsch nach nachhaltigen Produkten findet sich dann unter „sozial“ und „ökologisch“. In der praktischen Anwendung der Bewertung hat sich bewährt, vor der Bewertung ein gemeinsames Verständnis zu schaffen und zu dokumentieren, was genau mit den Clustern in Bezug auf das Produkt gemeint ist.

3 Zusammenfassung

In der Arbeit wurde gezeigt, wie Nachhaltigkeitskriterien und Nutzerzentrierung in die Bewertung mit TRIZ-Werkzeugen einfließen können. Da Nachhaltigkeit und Nutzerzentrierung häufig mit höheren Kosten einhergehen, ist es nun möglich, eine quantitative Vergleichbarkeit herzustellen.

Die Funktionsbewertung, die in die „Wertanalytische Betrachtung" [1] einfließt, schafft Transparenz und die Möglichkeit, Aufgaben für die Optimierung abzuleiten.

Mit der Möglichkeit, die Idealität verschiedener Konzepte quantifizierbar zu machen, wird eine Entscheidungsfindung erleichtert und zusätzlich können für Projekte verschiedene Szenarien mithilfe unterschiedlicher Kriterien durchgespielt werden.

Quellen

1. Adunka, R.: „TRIZ Level 1" Trainingsunterlagen. TRIZ Consulting Group GmbH, Nürnberg (2014)
2. LIT I Ahrend, K.-M.: Geschäftsmodell Nachhaltigkeit: ökologische und soziale Innovationen als unternehmerische Chance. Springer, Berlin (2019)
3. LIT II Bocken, N.M.P., Rana, P., Short, S.W.: Value mapping for sustainable business thinking. J. Ind. Prod. Eng. **32**(1), 67–81 (2015)
4. Ceschin, F., Gaziulusoy, I.: Evolution of design for sustainability: from product design to design for system innovations and transitions. Des. Stud. **47**, 118–163 (2016)
5. Corsini, F., et al.: Consumer's Circular Behaviors in Relation to the Purchase, Extension of Life and End of Life Management of Electrical and Electronic Products: A Review. www.mdpi.com/journal/sustainability (2020)
6. Iniotakis, C.: VA++-The Next Generation of Value Analysis in TRIZ. MATRIZ Official: TRIZ Future Conference (2022)
7. Lyubomirskiy, A., et al.: Trends of Engineering System Evolution. TRIZ Consulting Group, Sulzbach-Rosenberg (2018)
8. Mann, D.: Business Matrix 3.0, 1. Aufl. IFR Press, United Kingdom (2018)
9. Mayer, Oliver. Sustainability modelling with TRIZ. TRIZfest-2022 170 (2022)
10. Mignoli, J.: Innovative Werkzeuge zur Kosteneinsparung durch den Einsatz von TRIZ. TRIZ-Expertentag, Sulzbach-Rosenberg (2024)
11. Orso, Mignoli, Gerundt: „Vorgehensmodell zur Kostenanalyse unter Verwendung der Funktionsanalyse nach TRIZ. TRIZ-Anwendertag, bayern-innovativ; Nürnberg (2024)
12. Schomaker, R.M., Sitter, A.: Die PESTEL-Analyse–Status quo und innovative Anpassungen. Der Betriebswirt 61. Jahrgang, Heft 1, S. 9–27. Duncker & Humblot, Berlin (2020)
13. Russo, D., Spreafico, C.: TRIZ-based guidelines for eco-improvement. Sustainability **12**(8), 3412 (2020)
14. Spreafico, C.: Quantifying the advantages of TRIZ in sustainability through life cycle assessment. J. Clean. Prod. **303**, 126955 (2021)
15. VDI 4521 Blatt 1

Vorgehensmodell zur Kostenanalyse unter Verwendung der Funktionsanalyse nach TRIZ

Jochen Orso[1], Justin Mignoli[1] und Oliver Gerundt[2](✉)

[1] Hochschule Reutlingen, Reutlingen, Deutschland
{jochen.orso,justin.mignoli}@reutlingen-university.de
[2] Robert Bosch GmbH, Gerlingen-Schillerhöhe, Deutschland
oliver.gerundt@de.bosch.com

Zusammenfassung. Vorgestellt wird ein Vorgehensmodell zur Analyse der Kostensituation von Systemen und Komponenten. Beschrieben wird zunächst der aktuelle Status quo als Basis, um Lösungen zur Kostenreduzierung von Systemen und Produkten zu erarbeiten. Hierzu sind im Wesentlichen die wertanalytische Betrachtung nach TRIZ und die Funktionskostenanalyse aus dem Bereich Product Engineering bzw. aus der VDI-Norm verbreitet, die jeweils spezifische Vor- und Nachteile mit sich bringen. Im hier vorgeschlagenen Modell werden die jeweiligen Vorteile zu einem neuen Vorgehen kombiniert.

Schlüsselwörter: TRIZ · Wertanalytische Betrachtung · Funktionskostenanalyse · Rationalisierung · Feature Transfer

1 Ausgangssituation

Der Kostendruck bei der Herstellung von Produkten steigt. Insbesondere bei Produkten, die schon länger im Markt sind und bei denen starke Wettbewerber vorhanden sind. Im Wesentlichen setzen sich die Produktkosten bei der Herstellung aus den Materialkosten, der Wertschöpfung und den Gemeinkosten zusammen. Die Gemeinkosten wie Vertrieb und andere, werden an dieser Stelle nicht weiter betrachtet. Die Materialkosten sind die Kosten des eigentlichen Materials aber auch die Kosten von Bauteilen, die von einem Zulieferer angeliefert werden. Unter Wertschöpfung versteht man die Kosten, die anfallen, um das Bauteil im Produkt zu verbauen bzw. zu montieren. Diese Kosten werden von der Entwicklung, dem Produktdesign und der Fertigung definiert.

Der Prozess der etablierten Kostenanalyse sieht meist so aus, dass sich eine Expertengruppe aus Entwicklung, Fertigung und Einkauf trifft und die teuersten Bauteile durchspricht, wobei Maßnahmen zur Kosteneinsparung besprochen werden. Das Ziel der hier vorgestellten Methoden zur Kostenanalyse ist die Visualisierung der Ausgangssituation und beinhaltet noch keine Werkzeuge zur Lösungsfindung. Die Aufgaben zur

O. Mayer (Hrsg.): TRIZ-Anwendertag 2024,
https://doi.org/10.1007/978-3-662-70698-5_11

Lösungsfindung beziehungsweise Ideengenerierung können aber aus den Ergebnissen der Analysearbeit abgeleitet werden.

Im Folgenden werden zwei etablierte Methoden zur Analyse der Kostensituation beschrieben, sowie eine verbesserte Methode, die exaktere und in der Praxis nützlichere Ergebnisse liefert vorgestellt. Eine heute weit bekannte aber nicht immer genutzte Methode ist die „Funktionskostenanalyse", wie sie im Rahmen der VDI 2800 – „Wertanalyse" verwendet wird (Nachfolgend als „VDI" bezeichnet). Eine weitere Methode ist die sogenannte „Wertanalytische Betrachtung" nach TRIZ. Diese ist im industriellen Kontext weniger bekannt.

Aus diesen beiden speziellen Methoden wurde im Rahmen dieser Arbeit eine dritte abgeleitet, welche die spezifischen Vorteile miteinander kombiniert und als neuer Standard vorgeschlagen wird.

Beispiel:
Um die Erkenntnisse konkreter darstellen zu können, werden diese nachfolgend anhand eines Beispiels erklärt. Hierbei wurde ein sehr einfaches System gewählt, ein Schlüsselanhänger in Form eines Flaschenöffners mit Werbeaufdruck (Abb. 1).

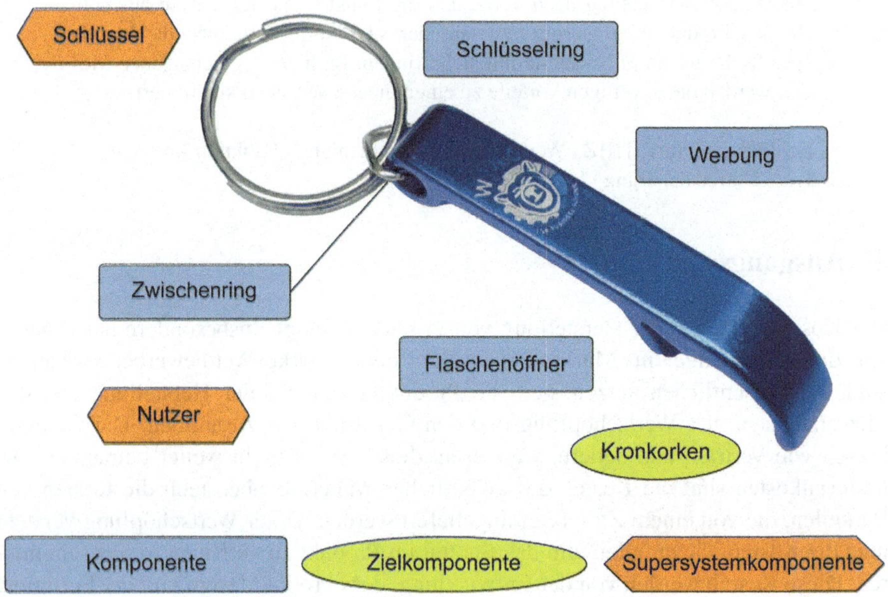

Abb. 1: Beispielsystem: Flaschenöffner-Schlüsselanhänger mit Werbeaufdruck

1.1 Funktionskostenanalyse nach VDI

Die Funktionskostenanalyse ist ein Analysewerkzeug, um Kosteneinsparprojekte systematisch zu unterstützen. Hier wird dargestellt, wie sie im Rahmen der Wertanalyse nach VDI 2800 verwendet wird. Sie beschreibt die Funktionen auf Basis der Funktionsanalyse, wie sie in Pahl/Beitz [2]. Diese Methode definiert Funktionen als den gewollten Zusammenhang zwischen Input und Output eines Systems, um die gewollte Aufgabe zu erfüllen. Die Definition betont die Gestaltung von Systemprozessen zur Zielerreichung und hebt die lösungsneutrale Darstellung der Funktionen hervor. Funktionen dienen dabei als Anforderungsträger und umfassen die Transformation beziehungsweise den Fluss von Stoffen, Energie und Signalen.

Die Funktionen werden anschließend nach ihrem Beitrag an der Gesamtfunktion gewichtet und dann der Beitrag der einzelnen Systemkomponenten an den Funktionen bestimmt.

Durch die Verrechnung der Funktionsgewichtung mit den anteiligen Beiträgen der Komponenten an den Funktionen, können zwei Diagramme, wie in Abb. 2 abgeleitet werden.

Im oberen Diagramm ist die Komponentensicht gezeigt. Auf der x-Achse sind die Zielkosten aufgetragen. Die Zielkosten sind die Kosten, die eine Komponente aufgrund ihres Beitrags zur Funktionserfüllung des Produkts und den Zielkosten des Produkts, haben dürfte. Auf der y-Achse sind die Istkosten der Komponenten aufgetragen. Ziel der Kostenarbeit ist, dass alle Einzelkomponenten unterhalb der Winkelhalbierenden sind, wo die Istkosten der Komponenten kleiner als die Zielkosten sind.

Im unteren Diagramm sind analog dazu die Funktionen dargestellt. Insbesondere in einer sehr frühen Phase eines Projekts, in dem noch nicht alle Komponenten klar sind, bietet diese Darstellung Vorteile.

Ziele der Funktionskostenanalyse sind:

- Kosten und Kostenstruktur für ein Produkt oder System transparent darzustellen
- Kosten für Spezifikationen bzw. Anforderungen abzuleiten
- Vergleich verschiedener Konzepte
- Der Abgleich der Ist- oder Prognose-Kosten zu den Zielkosten

1.2 Wertanalytische Betrachtung nach TRIZ

Basis der „Wertanalytischen Betrachtung“ nach TRIZ ist die Funktionsanalyse des technischen Systems [7–9]. Im Unterschied zu VDI ist eine Funktion in TRIZ definiert als Interaktion zwischen zwei Komponenten, wobei ein Parameter des Objekts der Funktion verändert oder aktiv erhalten wird. Da in der Analyse die einzelnen Komponenten dargestellt werden, kann mit dem Modell konkreter gearbeitet werden als mit dem VDI-Funktionsmodell.

Das gezeigte Funktionsmodell zeigt den Zeitpunkt für die Hauptfunktion aus Sicht des Nutzers, das Öffnen einer Flasche. Die Sicht des Werbetreibenden wird an dieser Stelle nicht betrachtet (Abb. 3).

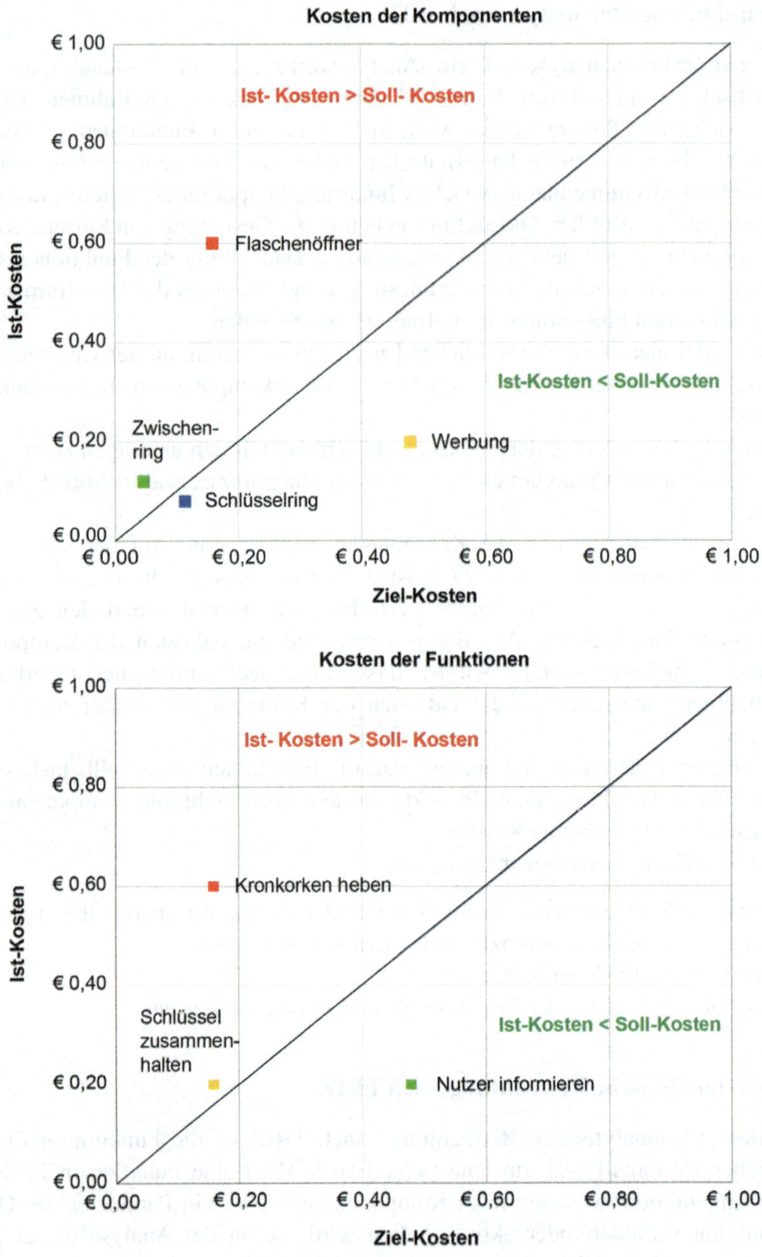

Abb. 2: Diagramme der Funktionskostenanalyse nach VDI

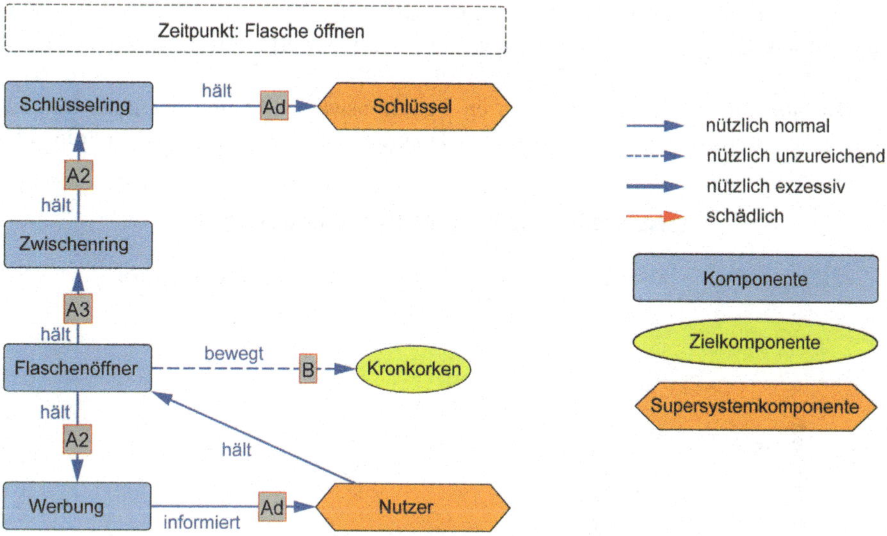

Abb. 3: Funktionskostenanalyse am Beispiel Schlüsselanhänger

Im Anschluss an die Funktionsmodellierung erfolgt die Ermittlung des Rangs einer jeden Funktion nach festen Regeln (B; A1, A2… An; Ad), wobei B die Basisfunktion, A1 bis An die Hilfsfunktionen und Ad die Zusatzfunktionen (Additional) bezeichnen. Begonnen wird mit der Funktion, die auf die Zielkomponente wirkt und als Basisfunktion bezeichnet wird, welche die wesentliche Hauptfunktion des technischen Systems darstellt. Die Zielkomponente ist in diesem Fall die Komponente, auf die das System angewendet wird. Die Zusatzfunktionen wirken auf die sogenannten Supersystemkomponenten, die sich außerhalb des Systems befinden bzw. es umgeben. Anschließend werden die Werte der internen Funktionen von der Basisfunktion aus vergeben. Je weiter entfernt von der Basisfunktion, steigt der Index der Funktionsränge und sinkt der spätere Funktionswert.

Aus dem Beispiel ergeben sich somit folgende Funktionswerte für die Funktionsränge (Tab. 1):

Tab. 1: Zuordnung Funktionswerte zu Funktionsrängen

Funktionsrang	Funktionswert
B	5
A1 = Ad	3
A2	2
A3	1

Auf den Flaschenöffner durch drei Funktionen mit den Werten 5, 2 und 1, ergibt sich ein Gesamt-Funktionswert von 8. Da das der höchste Wert gesamt ist, müssen alle Werte zur Normierung durch 8 geteilt werden. Damit ergibt sich für die Werbung ein Gesamt-Funktionswert von 3 – normiert 3/8 = 0,375. Für den Schlüsselring und den Zwischenring ergeben sich die Gesamt-Funktionswerte von 2 – normiert 2/8 = 0,25.

Die auf den Höchstwert normierten Funktionswerte mit den ebenfalls auf den Höchstwert normierten Kosten in einem Diagramm aufgetragen, entsteht folgendes

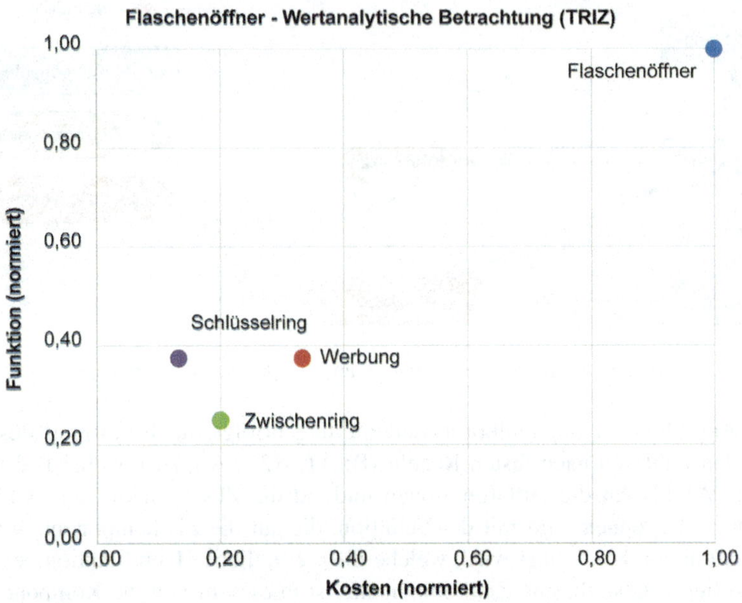

Abb. 4: Stärkediagramm für wertanalytische Betrachtung nach TRIZ

Diagramm (Abb. 4):

Im Diagramm ist zu erkennen, dass der Flaschenöffner den höchsten Funktionswert und gleichzeitig den höchsten Kostenanteil in sich vereint. Dabei haben der Zwischenring und der Schlüsselring einen geringeren Funktionsumfang aber auch einen geringen Kostenanteil.

Für den Werbeartikel ist im vorliegenden Beispiel der Funktionswert des Werbeaufdrucks stark unterbewertet. Aus Sicht der werbenden und bezahlenden Firma ist die Werbung die eigentliche Hauptfunktion. Aus diesem Beispiel, sowie auch schon in anderen Veröffentlichungen [3, 4] gezeigt entsteht der Wunsch nach einer individualisierbaren Vergabe der Funktionswerte.

1.3 Gegenüberstellung der etablierten Methoden

Die beiden vorgestellten, etablierten Methoden bieten jeweils bestimmte Vor- und Nachteile bei der Anwendung und auch bei der Nutzbarkeit der Ergebnisse.

Die Vorteile der Funktionskostenanalyse nach VDI:
Da die Beschreibung der Funktionen auf der allgemeinen Funktionsbeschreibung nach VDI basiert, ist diese insbesondere in den frühen Phasen der Wertgestaltung geeignet, wo noch nicht alle konstruktiven Details definiert sind. Eine echte Stärke der Methode ist jedoch, dass die Gewichtung der Funktionen nach ihrer Wichtigkeit bzw. der Höhe des Beitrags zur Gesamtfunktion individuell erfolgt.

Die Nachteile der Funktionskostenanalyse nach VDI:
Da die Funktionen des Systems allgemeingültig beschrieben werden, sind die Diskussionen über die Funktionen von den Bauteilen entkoppelt. Für die Erstellung der Übersichtsdiagramme werden die Bauteilanteile an den Funktionen individuell gewichtet, wodurch zwei Bauchgefühlbewertungen miteinander multipliziert werden, was wiederum zu größerer Ungenauigkeit führt. Die Darstellungen der Ergebnisse zeigen den Zusammenhang der Ziel- und Istkosten der Komponenten und der Bauteile unabhängig voneinander. Eine Darstellung echter Funktionskosten gibt es hier nicht.

Die Vorteile der „Wertanalytischen Betrachtung“ nach TRIZ
Die Funktionsanalyse nach TRIZ, auf der die „Wertanalytische Betrachtung“ basiert, hat den Vorteil, dass alle Bauteile des Systems als Komponenten in der Analyse modelliert werden. Somit entsteht ein grafisches Funktionsmodell, bei dem alle Interaktionen sichtbar sind. Im Falle von Änderungen an den Schnittstellen sind die Auswirkungen auf benachbarte Bauteile direkt ableitbar. Die Darstellung im Diagramm zeigt die Wertigkeit der Funktion eines jeden Bauteils über deren Kosten, wodurch man hier von einer echten Funktionskostenanalyse sprechen kann.

Die Nachteile der „Wertanalytischen Betrachtung“ nach TRIZ
Die Beurteilung der Wertigkeit der Funktionen erfolgt durch klare Regeln. Dieser große Vorteil bringt jedoch auch eine fehlende Flexibilität mit sich. Insbesondere bei Systemen, die einen stark linearen Aufbau haben und Komponenten weit entfernt von der Zielkomponente eine hohe Funktionalität besitzen, führt das zu Verzerrungen der Realität. Dieser Fall tritt meist bei Systemen mit intelligenter Steuerungselektronik auf, sodass diese stark unterbewertet wird. Verzerrungen der Funktionswerte kommen erfahrungsgemäß auch bei Funktionen vor, die die Nutzerzentrierung und Kundenbegeisterung betreffen. Diese „soften“ Funktionen sind in der klassischen „Wertanalytischen Betrachtung“ meist unterbewertet (Tab. 2).

Die Verbesserung der etablierten Methoden wurde unter anderem schon von C. Iniotakis angegangen und als VA++ veröffentlicht [3].

2 Beschreibung des Vorgehensmodells

Die Vorgehensweise nach VDI und TRIZ besitzen an unterschiedlichen Stellen Vor- und Nachteile. Das TRIZ-Werkzeug „Feature Transfer“ setzt an dieser Stelle an und überträgt die Vorteile des einen Systems auf die Stelle des anderen Systems, bei dem dieses Nachteile besitzt (Tab. 3).

Tab. 2: Vor- & Nachteile der TRIZ- und VDI-Vorgehensweise

TRIZ		VDI	
+/–	Bewertung und Gewichtung der Funktionalität durch systematische Zuweisung des funktionalen Werts	–	Bewertung und Gewichtung der Funktionalität durch Bauchgefühl/Diskussionen
–	Kann zu Verzerrungen der Realität führen		
+	Keine Teamdiskussion während der Durchführung der Analyse	+/–	Die Funktionen werden individuell gewichtet und anschließend individuell zu den Komponenten bewertet
–	Keine Nutzung von Zielkosten	+	Gegenüberstellung von Istkosten und Zielkosten

Tab. 3: Feature Transfer [1]

	TRIZ „Wertanalytische Betrachtung“	VDI „Funktionskostenanalyse“
Funktionsdarstellung	+	–
Integration Zielkosten	–	+

Die Zielsetzung des neuen Vorgehensmodells besteht darin, die Methodik der Funktionsdarstellung gemäß TRIZ zu nutzen und gleichzeitig die Zielkosten so zu integrieren, dass sowohl Projekte im Allgemeinen als auch speziell die Einkaufsabteilung effektiv damit arbeiten können.

Somit greift das vorgeschlagene Vorgehensmodell auf die lösungsorientierte Funktionsanalyse gemäß TRIZ zurück. Im Unterschied dazu werden die Funktionswerte jedoch individuell basierend auf festgelegten Kriterien vergeben. Zudem werden, ähnlich der Funktionskostenanalyse, die Zielkosten in die Übersicht integriert. Dadurch können sie mit den Istkosten verglichen werden, was die Identifizierung von Schwerpunkten für die weitere Bearbeitung erleichtert.

Ursprünglich wurde das Vorgehensmodell entwickelt, um die Analysephase in Projekten zur Kosteneinsparung zu verbessern. Daher erhielt es den Namen „Value Analysis for Ratio Projects, V4RP“. Es ist jedoch nicht nur auf die Anwendung in Ratioprojekten beschränkt, sondern kann auch während der Wertgestaltungsphase in der Entwicklungsphase genutzt werden.

2.1 Funktionsanalyse

Der erste Schritt von V4RP ist die Funktionsanalyse, wie oben in Abb. 3 beschrieben, jedoch ohne Vergabe der Funktionsränge (B, A1–An, Ad).

2.2 Funktionsbewertung

Die Wertigkeit der Funktionen soll individuelle vergeben werden. Um die Vergabe der Werte zu erleichtern und die Bewertung nachvollziehbar zu dokumentieren, wurde die

Methode des paarweisen Vergleichs verwendet. Hierbei wird jede Funktion mit allen anderen Funktionen verglichen und entschieden, ob die eine oder andere Funktion wichtiger oder beide gleich wichtig sind.

- Eine Funktion wird mit 1 bewertet, wenn sie gleichwertig mit ihrer Vergleichsfunktion ist.
- Eine Funktion wird mit 2 bewertet, wenn sie wertiger als ihre Vergleichsfunktion ist.
- Eine Funktion wird mit 0 bewertet, wenn sie minderwertiger als ihre Vergleichsfunktion ist.

Die folgende Tabelle zeigt die Vergleichstabelle anhand unseres Beispiels (Tab. 4).

Tab. 4: Paarweiser Vergleich für Schlüsselanhänger

	Flaschenöffner bewegt Kronkorken	Flaschenöffner hält Werbung	Flaschenöffner hält Zwischenring	Werbung informiert Nutzer	Zwischenring hält Schlüsselring	Schlüsselring hält Schlüssel	**Summe Funktionalität**	**Beitrag an die Gesamtfunktion**	**Funktionalität normiert**
Flaschenöffner bewegt Kronkorken		2	2	0	2	2	**8**	**27%**	**8**
Flaschenöffner hält Werbung	0		2	0	2	2	**6**	**20%**	**6**
Flaschenöffner hält Zwischenring	0	0		0	1	1	**2**	**7%**	**2**
Werbung informiert Nutzer	2	2	2		2	2	**10**	**33%**	**10**
Zwischenring hält Schlüsselring	0	0	1	0		1	**2**	**7%**	**2**
Schlüsselring hält Schlüssel	0	0	1	0	1		**2**	**7%**	**2**

Die Leitgedanken für die getroffene Bewertung des paarweisen Vergleichs sind wie folgt beschrieben:

- Werbung informiert Nutzer…
  - …stellt für den Werbetreibenden, der das Produkt finanziert, die wertigste Funktion dar. Ohne Werbung würde das Produkt nicht produziert werden.
- Flaschenöffner bewegt Kronkorken…
  - … ist die Hauptfunktion für den Nutzer und somit wertiger als alle Haltefunktionen.
  - … ist weniger wertig als Werbung informiert Nutzer, da ohne diese Funktion der Werbetreibende das Produkt nicht herstellen würde
- Flaschenöffner hält Werbung…

  - … ist aufgrund der Nähe Hauptnutzen wertiger als die internen Haltefunktionen.
  - … ist weniger wertig als die informierende Funktion der Werbung (siehe oben).
- Flaschenöffner hält Zwischenring, Zwischenring hält Schlüsselring, Schlüsselring hält Schlüssel…
  - … sind alle gleichwertig, da es sich um gleichartige interne Haltefunktionen handelt

Die Entscheidung, welche Funktion bei einem Vergleich wichtiger ist, ist oft nicht eindeutig und führt zu Diskussionen. Um dies zu erleichtern, wurde ein Kriterienkatalog erarbeitet, der dies erleichtern soll. Durch die Vielfalt an Systemen ist es schwer einen einzigen Kriterienkatalog für alle Systeme zu entwickeln. In der Praxis macht es Sinn, für das zu bewertende System, individuelle Kriterien zu erstellen. Diese sind abhängig vom Produkt, dem Markt, den Kunden usw.

Tab. 4 zeigt einen Ausschnitt der wesentlichen Kriterien, die sich in der Praxis bewährt haben (Tab. 5).

Im Anschluss an den paarweisen Vergleich, werden die Funktionswerte in eine Tabelle übertragen. Dabei werden die Funktionen mit dem gleichen Funktionsträger kumuliert und je Komponente addiert. Anschließend werden die Werte auf den höchsten Einzelwert normiert. Mit den Kosten wird gleich verfahren. Die Kosten werden je Komponente ermittelt und ebenfalls auf den höchsten Wert normiert (Tab. 6).

Werden die normierten Funktionswerte und die Kosten in ein Diagramm übertragen, so ergibt sich folgende Abb. 5:

Im Vergleich zu TRIZ, wird hier erkennbar, dass sich an dem Flaschenöffner nichts geändert hat. Im Vergleich zur wertanalytischen Betrachtung nach TRIZ, hat sich die Bedeutung der Werbung verändert. Der Zwischenring und der Schlüsselring besitzen hier im Verhältnis weniger Wert.

Somit wurde gezeigt, dass die ursprüngliche Intention, eine realistische Bewertung der Funktionen über die Kosten zu erhalten, erreicht wurden.

Die Analyse des vorhandenen Systems hat das Ziel, die richtigen Arbeitsschwerpunkte abzuleiten. Hierzu ist es noch hilfreich zu wissen, wie teuer die einzelnen Komponenten aufgrund ihres Funktionsbeitrags sein dürften, so wie dies bei der VDI-Funktionskostenanalyse der Fall ist.

Die nachfolgende Tabelle zeigt, wie diese Integration umgesetzt wird (Tab. 7).

Die Zielkosten der Komponente sind der Anteil an den Gesamtzielkosten verhältnismäßig des Funktionsanteils der Komponente an der Gesamtfunktion.

Die Zielkosten des Systems werden vom Vertrieb auf Basis des Marktes festgelegt.

Sind die Zielkosten geringer als die Istkosten ergibt sich daraus einen Handlungsschwerpunkt (rot dargestellt) liegen die Zielkosten über den Istkosten, so haben diese einen geringeren Handlungsbedarf (grün dargestellt).

Eine weitere Möglichkeit der Kostenoptimierung bildet die funktionale Betrachtungsweise. Hier werden die Ist- und Zielkosten der Funktionen verglichen.

Die abgeleiteten Arbeitsfragen sind im ersten Fall, wie können die einzelnen Komponenten günstiger dargestellt werden und im zweiten Fall wie können die einzelnen Funktionen günstiger dargestellt werden?

Tab. 5: Kriterienkatalog für paarweisen Vergleich (PWV) von Funktionen

Bewertung	Funktion	Kriterien
Wertigere Funktionen (PWV mit 2)	Basisfunktion	Die Basisfunktion wirkt direkt auf die Zielkomponente und hat somit im Normalfall die höchste Priorität
	Beitrag zur Hauptfunktion	Funktionen, die einen direkten Beitrag zur Hauptfunktion des Systems haben, werden sehr hoch bewertet Z. B. können alle Übertragungselemente in einem Antriebsstrang die gleiche Priorität bekommen
	Funktionsautonomie	Kann die Komponente diese Funktion selber erfüllen oder braucht sie Unterstützung einer nachgeordneten Komponente? Z. B. kann die Komponente, die direkt auf die Zielkomponente wirkt, nur ein Übertragungselement sein und die eigentlich wertige Funktion aus einer Komponente kommen, die geometrisch weiter weg angeordnet ist
	Erhöhte Anforderungen	Werden erhöhte Anforderungen an eine Funktion gestellt, kann eine höhere Gewichtung zugewiesen werden Hierzu können Festigkeit bei mechanischen oder Rechenleistung bei mechatronischen Systemen zählen
	Hauptnutzungszeit	Funktionen, die in der Zeit auftreten, in der die Hauptfunktion des Produkts ausgeführt wird, sind wichtiger als Funktionen, die in der Nebenzeit ausgeführt werden. Dazu zählen beispielsweise Ladefunktionen
	Zusatzfunktionen	Zusatzfunktionen sind Funktionen, die auf andere Supersystemkomponenten wirken. Sie sind gegenüber Haltefunktionen zu priorisieren
Gleichwertige Funktionen (PWV mit 1)	Gleiche Funktionsart	Gleichartige Funktionen können als gleichwertig bewertet werden, unabhängig von der geometrischen Position im System Dazu zählen z. B. Haltefunktionen für Bauteile im System
	Ähnliche Funktionsart	Zwei Funktionen können gleich bewertet werden, wenn kein Priorisierungskriterium gefunden werden kann, dass eine Entscheidung zulässt Das kommt häufig bei untergeordneten Hilfsfunktionen vor
Weniger wertige Funktionen (PWV mit 0)	Untergeordnete Hilfsfunktionen	Hilfsfunktionen werden meist untergeordnet, sofern sie keinen Beitrag zur Hauptfunktion haben

Tab. 6: Wertetabelle

Funktionsträger	Funktionen	Funktions-wert (V4RP)	Σ Funktions-werte	Funktions-wert normiert/ Funktion-alität	Σ Ist-Kosten	Kosten normiert
Flaschenöffner	bewegt Kronkorken	8	16	1,00	0,60 €	1,00
	hält Werbung	6				
	hält Zwischenring	2				
Werbung	informiert Nutzer	10	10	0,63	0,20 €	0,33
Zwischenring	hält Schlüsselring	2	2	0,13	0,12 €	0,20
Schlüsselring	hält Schlüssel	2	2	0,13	0,08 €	0,13
SUMME			30		1,00 €	

Um eine gesamtheitliche Betrachtung zu schaffen, können die Ergebnisse aus dem paarweisen Vergleich, sowie aus der Tab. 5 ein Projekt-individuelles Cockpit Chart erarbeitet werden.

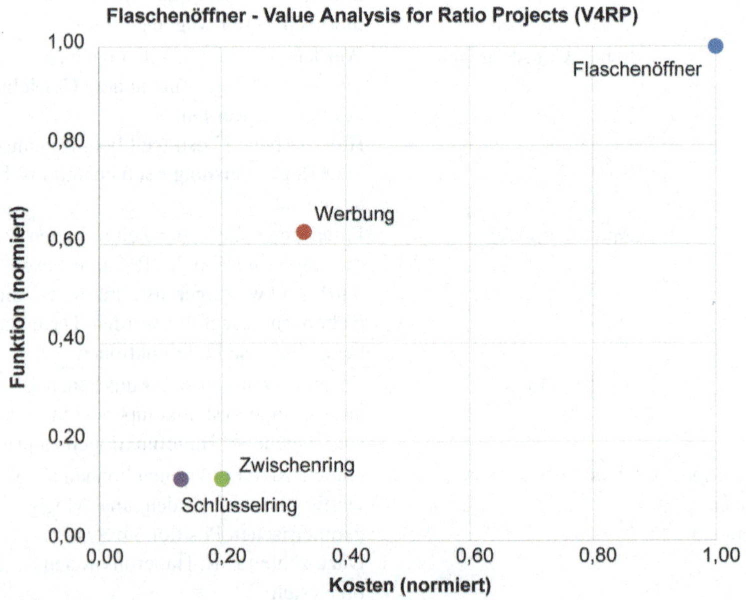

Abb. 5: Stärkediagramm für wertanalytische Betrachtung nach V4RP

3 Diskussion der neuen Methode

Die Methode bricht mit dem Vorgehen aus TRIZ, die Werte von Funktionen nach klaren Regeln zu vergeben. Die regelbasierte Vergabe bietet die Vorteile, dass keine Diskussionen erforderlich sind und diese Weise sehr zeiteffizient ist. Dem gegenüber steht der wesentliche Nachteil, dass bei vielen Systemen die Wertigkeit von Funktionen unter- oder überbewertet ist. Das ist insbesondere dann der Fall, wenn sehr wertige Komponenten, wie intelligente Steuerungen, tief im System integriert sind. Hier

Tab. 7: Wertetabelle mit integrierten Zielkosten

Funktionsbeitrag der Komponente x Zielkosten des Systems = Zielkosten der Komponente

Funktionsträger	Funktionen	Funktionsbeitrag innerhalb Komponenten-sicht (V4RP)	Gesamt-komponentenbeitrag zur Funktion	Funktions-wert normiert/ Funktion-alität	Σ Ist-Kosten	Ist-Kosten aufgeteilt auf die Funktionen	Σ Zielkosten	Zielkosten aufgeteilt auf die Funktionen
Flaschenöffner	bewegt Kronkorken	50%	53%	1,00	0,60 €	0,30	0,43 €	0,21
	hält Werbung	38%				0,23		0,16
	hält Zwischenring	13%				0,08		0,05
Werbung	informiert Nutzer	100%	33%	0,63	0,20 €	0,20	0,27 €	0,27
Zwischenring	hält Schlüsselring	100%	7%	0,13	0,12 €	0,12	0,05 €	0,05
Schlüsselring	hält Schlüssel	100%	7%	0,13	0,08 €	0,08	0,05 €	0,05
SUMME			100%		1,00 €	1,00 €	0,80 €	0,80 €

Funktionsbeitrag x Ist-Kosten = Ist-Kosten der Funktion

Funktionsbeitrag x Ziel-Kosten = Ziel-Kosten der Funktion

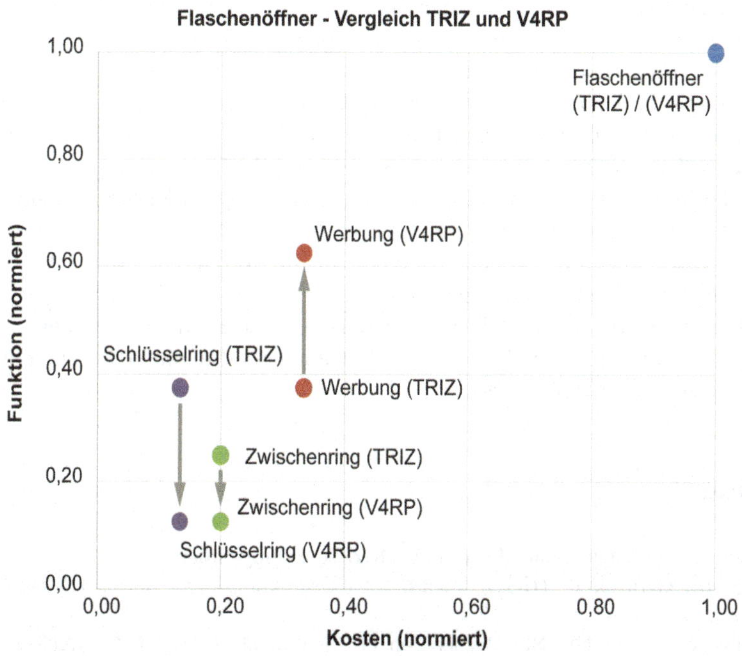

Abb. 6: Gegenüberstellung Darstellung TRIZ und V4RP

brauchte es bisher sehr spezielle Kenntnisse zu Kniffen in der Modellierung, um deren Wertigkeit erhöhen. Auch wenn, wie im oben gezeigten Beispiel, Funktionen eine wichtige Rolle spielen, die im Bereich der Interaktion mit dem Supersystem liegen, sind diese meist unterbewertet. Insbesondere bei Systemen und Produkten für Endkunden ist das der Fall, da diese auch als Supersystemkomponenten modelliert werden.

Dem Nachteil des zeitlichen Mehraufwands der neuen Methode wurde mit einem systematischen Vorgehen begegnet, um den Aufwand in Grenzen zu halten und auch um die Bewertung nachvollziehbar zu dokumentieren. Hierbei hat sich der paarweise Vergleich in der Praxis bewährt. Der dazugehörige Kriterienkatalog gibt Orientierung, sollte in der praktischen Anwendung aber nicht dogmatisch angewendet werden. Er ist eher als Hinweis zu verstehen und kann ergänzt oder verändert und mit Ausnahmen versehen werden. Mit der Branche können sich auch die Prioritäten ändern.

Abb. 6 zeigt in einem Diagramm die Veränderungen zwischen der klassischen Analyse nach TRIZ und der hier vorgeschlagenen Vorgehensweise V4RP.

Zu sehen ist, dass sich der Funktionswert der Werbung hin zu einem Wert verändert hat, der eher der Realität entspricht. Der Flaschenöffner dient als Werbeträger. Daher muss der Wert der Werbung auch entsprechend repräsentiert werden.

4 Zusammenfassung

Mit der neuen Methode der Kostenanalyse von Produkten konnte ein großer Schritt gemacht werden, um TRIZ in der Kostenarbeit im Unternehmenskontext zu etablieren. Die Vorgehensweise und Dokumentation der Ergebnisse entsprechen der Domänenlogik im kaufmännischen Umfeld. Insbesondere sei hier die Integration der Zielkosten genannt. Auch dem Umstand, dass es verschiedene Stakeholder mit unterschiedlichen Interessen bei einem Produkt gibt, wird Rechnung getragen.

Um die Methode weiter zu verbessern ist es sinnvoll den Kriterienkatalog weiterzuentwickeln und Erfahrungen damit zu sammeln.

Der Schwerpunkt für die neue Methode ist die Wertverbesserung, das heißt die regelmäßig stattfindende Kostenarbeit.

Im Bereich Wertgestaltung, insbesondere in der frühen Phase noch ohne konkretes System oder aber reinen Softwareprodukten, kann weiterhin die klassische Funktionskostenanalyse von Nutzen sein [5, 6].

Literatur

1. Adunka, R.: Trainings Manual Level 1–3. TRIZ Consulting Group.
2. Bender, B., Gericke, K. (Hrsg.): Pahl/Beitz Konstruktionslehre. Springer Vieweg, Berlin (2021)
3. Iniotakis, C.: VA++-The Next Generation of Value Analysis in TRIZ. MATRIZ Official: TRIZ Future Conference (2022)
4. Iniotakis, C.: Functional Value of TRIZ Components – Universal Concept and Direct Implications. MATRIZ Official: TRIZ Future Conference (2023)
5. Lyubomirsky, et al.: Trends of Engineering System Evolution (TESE). TRIZ Consulting Group (2018)
6. Mayer, O.: Increased Addressing of Human Senses as a Trend. MATRIZ Organization for TRIZ L5 Application (2017)
7. VDI 2800 Blatt 1: Wertanalyse. Verein Deutscher Ingenieure e. V. (2010)
8. VDI 2801 Blatt 1: Wertanalytiker/Value-Manager/Wertanalytikerin/Value-Managerin Berufsbild. Verein Deutscher Ingenieure e. V. (2010)
9. VDI 2803 Blatt 1: Funktionenanalyse Grundlagen und Methode. Verein Deutscher Ingenieure e. V. (2019)

A Case Study of Collaborative Practice in Human-Centered Design (HCD) and Theory of Inventive Problem Solving (TRIZ)

Christoph Milz[1] und Uwe Schaumann[2](✉)

[1] CP-Consulting, Fairfield, CA, USA
christoph.milz@gmail.com
[2] Oberderdingen, Deutschland
u-u.schaumann@t-online.de

Abstract. This paper describes a collaborative case study that incorporates Human-Centered Design (HCD) and the Theory of inventive problem solving (TRIZ) and demonstrates the results and value generation possible when these two methods come together. Recognizing the complementary nature of HCD and TRIZ, the study focuses on a practical, low-complexity application that maintains the integrity of both methods. The project uses HCD to explore human-centered aspects, while TRIZ focuses on technical solutions. The systematic collaboration enables the integration of human-centered insights and technical problem solving, with a focus on systems thinking, divergence and convergence processes and a shared mindset of curiosity. Experience shows that such an approach not only fosters the creative process, but also produces solutions that are more comprehensive and robust than those generated by the single method alone.

Keywords: HCD · TRIZ · Collaborative Work

1 Introduction

Creative and successful product solutions must integrate seamlessly within the systems they inhabit. To enhance success rates, it is beneficial to systematically incorporate diverse perspectives and disciplines [1–3].

However, the common reality of standard innovation and product development processes often reveals deficiencies. These are typically attributed to time constraints and a lack of profound experience with the necessary methodologies. Despite recognition of this problem and various efforts to address it, it remains rare to find innovation teams that effectively apply these methodologies in their daily work [2–5].

O. Mayer (Hrsg.): TRIZ-Anwendertag 2024,
https://doi.org/10.1007/978-3-662-70698-5_12

The authors explored whether a practicable approach exists to unite two highly complementary disciplines: HCD and TRIZ. In their review of previous efforts aimed at integrating these methodologies [6–12], they found that most attempts focused on merging the methodologies, introducing new terminology, and resequencing steps. However, these integrated approaches did not meet the authors' expectations, which centered on a low-complexity and practical application while preserving the integrity of existing methodologies and frameworks.

Consequently, the authors chose a collaborative approach over integration or steps with a siloed alternation. They selected a real-world innovation challenge, which was addressed through the combined expertise of both authors, each proficient in one of the disciplines. This collaboration of the human and technical perspective yielded a variety of innovative solutions, derived from a comprehensive exploration of both the problem and solution spaces.

2 Collaboration Between HCD and TRIZ

Fundamentally, HCD and TRIZ share several essential elements that provide a solid foundation for collaboration:

- **Systems Thinking:** Both methodologies emphasize analyzing and exploring system interdependencies and connections, as well as the ability to zoom in and out of the system to enhance understanding.
- **Process of Divergence and Convergence:** They incorporate a deliberate act of diverging before converging, allowing for thorough exploration, and fostering creative solution generation.
- **Curiosity:** An underlying mindset of curiosity drives both approaches, encouraging continuous inquiry and innovation

At the same time, there are elements of each methodology that are distinct yet synergistic:

- **HCD focuses on:**
  - Uncovering insights and reframing the challenge with those insights
  - Anchoring all efforts in specific human experiences and interactions
  - Collaborative development and inclusion of different perspectives is part of the process at every stage.
- **TRIZ focuses on:**
  - The abstraction of key elements to draw inspiration from well-documented, analogous technical solutions
  - A rigorous approach to simplifying and eliminating redundancies by focusing on the efficient use of existing resources

2.1 The Role of TRIZ in this Collaboration

In this project, TRIZ was strategically employed to address the technical aspects, focusing on identifying undesired effects and issues to enhance functionality during the problem-solving process. The primary emphasis was on the "Why?" and "How?"—questions critical to understanding and resolving these technical challenges. It is important to note that TRIZ was not used to define the initial problem related to system usage, nor was it employed to validate the solutions developed [13].

2.2 The Role of HCD in this Collaboration

HCD was strategically employed to identify unmet needs related to usability, use, and meaning. This methodology facilitated a comprehensive evaluation and consideration from a human-centered perspective, incorporating the viewpoints of all relevant stakeholders, including business partners. It also considered the underlying incentives driven by their business models.

Notably, HCD's iterative approach to eliciting specific responses from stakeholders proved particularly effective in mitigating unintended consequences, a crucial aspect of genuine systems thinking. This contrasts sharply with the TRIZ approach, which, by its nature, confines itself to the technical system and relies on previously recorded insights.

2.3 Flow of Information and Synergistic Collaboration

The initial input came from a manufacturer who described a specific pain point on a product component and some preliminary ideas for a way to solve it with a more advanced technical concept. The main overarching process and flow came from HCD and can be broken down into the following three main phases (see Fig. 1):

1) Initial Insights and Framing: HCD methodologies were instrumental in generating deeper system insights. Through the identification, observation, and interviewing of both obvious and non-obvious stakeholders, we developed a comprehensive understanding. The outcome included a detailed set of insights, open questions, stakeholder maps detailing interactions, interdependencies, belief structures, and environmental limitations.

2) Exploration, Iteration, Reframing, and Solution Paths: The innovation challenge was initially framed and presented to the TRIZ collaborator through open, inviting, and generative questions starting with "How might we…?". A full overview of the human-centered findings and anecdotes was also shared, fostering empathy, and enriching the collaborative understanding.

This set the foundation for continuous exchange and deeper exploration, allowing both contributors to pose new questions influenced by the respective perspectives. These interactions enabled each contributor to integrate new understandings into their ongoing processes seamlessly, without compromising their methodologies. The result from this phase was a broad spectrum of concepts that extended beyond the initial challenge. Further data and insights prompted us to reframe the original challenge, leading to the definition of solution paths that balanced technical feasibility with stakeholder needs, beliefs, and business models.

3) Prototyping, Feedback, Hand-off to Development Team: The final phase involved developing low-fidelity prototypes, a key step in the HCD process designed

to elicit further feedback before handing off to the development team. The objective at this stage was not to validate the prototypes but to provoke discussion and identify any unintended consequences. Interestingly, this phase did not yield additional insights that necessitated revisiting earlier phases of collaboration.

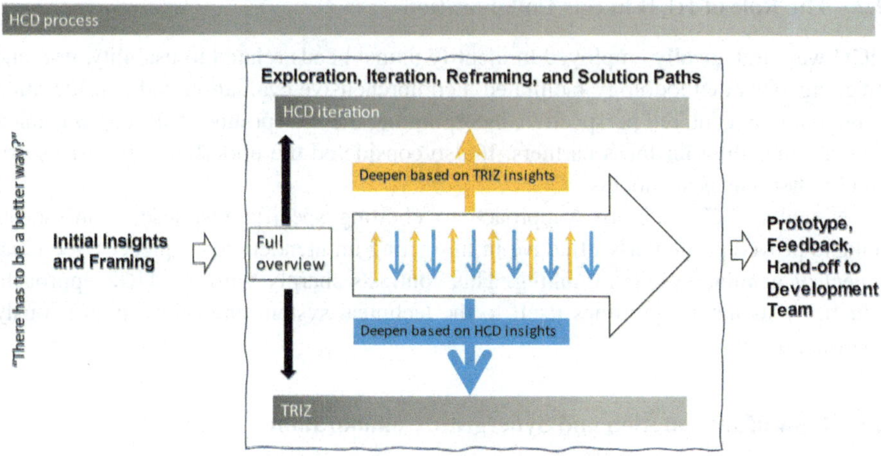

Fig. 1. Project planning and collaborative work in the analysis and solution finding phase.

2.4 Reframing and Identifying new Paths

The project underscored the importance of dynamic and iterative feedback loops from users and stakeholders impacted by the design. Through cycles of convergence and divergence, reflecting on these feedback loops, the authors saw the need and opportunity to reframe the initial challenge, moving the exploration outside of the narrow confines and another round of feedback loop from the users and stakeholders. This prevented the authors from arriving at narrow internal and discipline-driven perspectives, as it can often be a challenge as supported by the literature (2,7).

Notably, the principle of Ideality in TRIZ and the trial-and-error approach in HCD were both directed towards enhancing system value and user satisfaction, yet they diverged in execution, focusing on distinct aspects of the challenge (Fig. 2 and 3).

The Ideal Final Result (IFR) focuses on the ideal solution for the development of an existing technical system, utilizing the existing resources within the system and minimizing changes to the system or supersystem. This approach is consistent with short-term development tasks such as those in the case study [13,14,15]

Although the iteration of the HCD process requires adaptation of the IFR, this was not done in the case study to simplify the process. Instead, the HCD findings allowed for deviations from the ideal machine, which can be considered a step backwards. However, this deviation is justified by the iterations in the HCD process, which can be described as a new starting point. Therefore, it was not necessary to adjust the IFR, as shown in Fig. 3 with solutions 1 and 2, although they are outside the expected range using IFR for the technical solutions finding. This indicates that the solution found is adequate for the user, making further adjustment of the IFR unnecessary.

The ideality remains the criteria for the TRIZ solution process. In the collaborative work, however, it is used as a thinking model related only on the technical aspects.

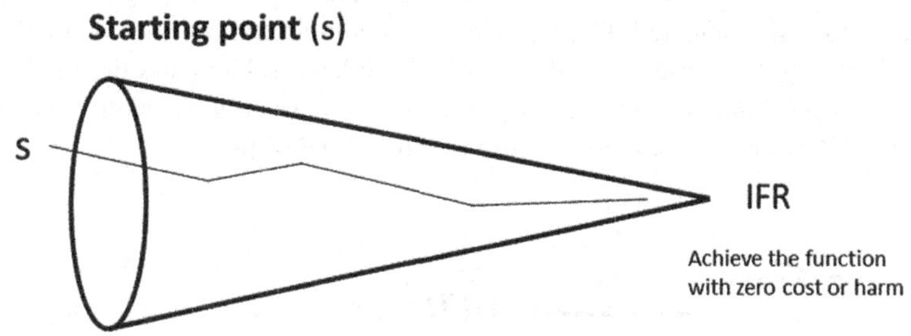

Fig. 2. Conical Search Space based on the Evolution of TRIZ Knowledge towards Ideal Final Result (IFR) [14].

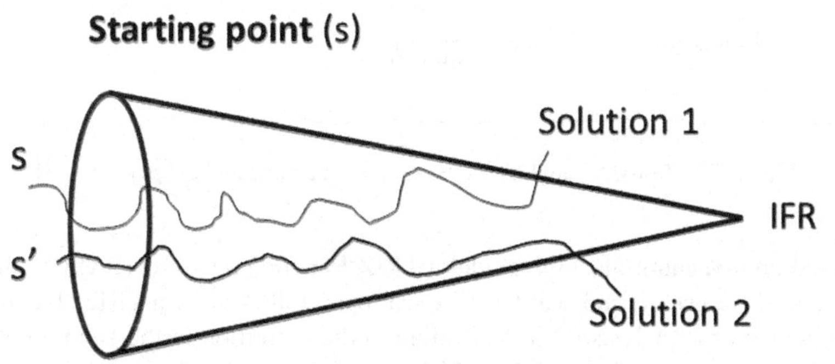

Fig. 3 Conical Search Space based on HCD iterations reframing the starting point, leading to solutions outside the search cone.

This collaborative dynamic reinforced the initial, intuitive decision to organize the project around HCD principles, while preserving the core values of both methodologies and emphasizing the integration of insights from parallel processes (Fig. 1).

3 Use Case

The system under investigation was a chamber vacuum packaging machine intended for semi-professional and commercial use. For the purposes of this description, a "generic" model representing current market solutions was employed (Fig. 4 top left). To simplify our discussion, we selected only one issue from a range identified through HCD. The challenge focused on enhancing both the usability and the lifespan of the seal bar, which is responsible for sealing specific plastic pouches through the application of heat and mechanical force.

Figure 4 shows the TRIZ tools used for the system and the seal bar analyses to generate insights and solutions shared with the HCD process (Fig. 1). The dotted arrow describes the selected TRIZ process. For simplification, the interaction with HCD is shown in the phase of the definition of the key problem and the final solution input. During the Function Analyses (FA) questions derived too which were given as input to the HCD process but did not stop the shown TRIZ process.

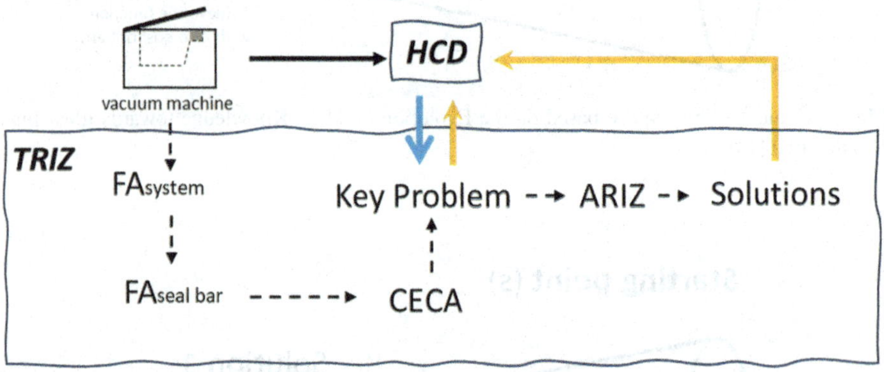

Fig. 4 TRIZ process applied on the vacuum machine and the seal bar [13].

Based on a vacuum machine model a TRIZ Function Analysis (FA) was done for the system (FA system) and seal bar (FA seal bar), followed by a TRIZ- based Cause and Effect Chains Analysis (CECA) leading to the definition of the key problem. For the problem-solving process, ARIZ, which is an acronym for the Russian phrase "Algorithm for Inventive Problem Solving" was selected. One specific improvement to the seal bar involved enhancing the consistency and distribution of the power supply, which in turn improved the sealing process towards an Ideal solution (see Fig. 2). During a collaborative exchange that considered unintended consequences, input from HCD on the overall context of the appliance and its installation provided a new starting point S' (Fig. 3). This led to an alternative path and solution endpoint 2 (Fig. 3), enhancing the value and flexibility of the solution to improve the power supply. This approach of reframing the starting point and discovering more impactful solutions occurred repeatedly.

The reframing and iteration depicted in Fig. 3 are typical of the HCD process and demonstrate the benefits of collaboration and seeking input from other contributors, in this case, a TRIZ partner.

From a TRIZ perspective, the new input and reframing provided by HCD did not disrupt the process and workflow to move towards the IFR. Instead, it highlighted certain blind spots in the understanding and presented interesting opportunities for re-design within the technical system increasing the input for the subsequent prototyping phase (Fig. 1), which is still in work.

4 Reflection and Summary

Based on the experience gained, the overall value created by the HCD-TRIZ collaboration can be summarized and used for further discussion. This applies also to single aspects like the solutions found and each method which were not found without a collaboration (Table 1).

Table 1. Values of the applied method.

Values of an HCD-TRIZ collaboration	
Enhanced Feedback Loops	The collaboration led to earlier and richer incorporation of system feedback loops involving technology, human factors, and context. This integration triggered iterative cycles of divergence and convergence, unveiling new potential paths
Rigorous Reframing	Multiple disciplinary perspectives enabled more rigorous reframing of the challenge throughout the process
Increased Concept Generation	The collaboration generated a greater number of concepts and specific ideas for solving the challenge than what would have been possible with a single-disciplinary approach alone
Value Transfer between Disciplines	Initially, HCD provided a more systematic briefing on the challenge's critical context than typical circumstances allow, enhancing both the speed and quality of TRIZ outputs
	In transitioning to higher fidelity prototyping and the development of technological solutions, TRIZ offered a more specific and targeted brief and references than HCD typically could, which increases the speed and lowers the cost of prototyping
Mutual Understanding	Each side gained a deeper appreciation of the other's methodologies

During the project, it was possible to determine factors that lead to a positive and creative work experience. The author believes for a collaboration between HCD and TRIZ methodologies to be successful, certain foundational mindsets are essential (Table 2).

Table 2. Factors and aspects for the Work Experience and needed mindset.

Factors for a positive and creative Work Experience	
Preservation of Disciplinary Integrity	The collaboration carefully avoided merging two distinct disciplines, thus maintaining the integrity and focus of each
Flexibility and Simplicity	The collaboration did not introduce any noticeable increase in complexity or time requirements, emphasizing an efficient and stimulating exchange
Ease of Integration	The process facilitated the inclusion of experienced practitioners from either discipline without necessitating additional training
Aspects for a shared mindset	
Curiosity in Perspectives	Participants must maintain a strong curiosity about each other's disciplinary perspectives. This curiosity fosters a deeper understanding and appreciation of the unique insights and approaches each discipline brings to the table
Openness to Feedback	Openness to receiving and integrating feedback is crucial. It allows for the continuous refinement of ideas and solutions, making the collaborative process more dynamic and effective
System Thinking	A robust practice of system thinking is vital. It enables participants to see beyond the immediate problem to understand the larger context in which the solution must operate. This understanding helps in crafting solutions that are not only innovative but also viable within the system's constraints and possibilities These mindsets not only facilitate smoother interactions but also enhance the overall quality and impact of the collaborative efforts. Embracing these attitudes ensures that both TRIZ and HCD methodologies can effectively converge to produce superior outcomes

While reflecting of the above topics the question arose as to how a collaborative structure could be further supported or done. Especially in the case of teams, it became clear from the discussion that the chosen approach and the form of exchange may reflect the concept of liberating structures. This idea also emerged during the final reflection and will need to be explored in greater depth in a further project.

5 Conclusion

As we embarked on this collaboration, the trajectory was initially uncertain. Concluding this project, this form of collaboration between TRIZ and HCD generated value that is surpassing the sum of its elements.

The initial decision to employ HCD to frame the overall project and plan initial iterations facilitated a smooth project start. It also ensured that the solutions comprehensively incorporated all crucial perspectives and necessary iterations from the outset.

In the core phase of collaboration for this project, both disciplines contributed significantly with complementing input to arrive at a more targeted understanding and more creative outcomes, as well as enhancing overall efficiency without extending the usual project timeline. The project demonstrated the feasibility of rapidly entering new fields, delivering prototypes, and generating deep, insightful exchanges with experts, non-obvious stakeholders, and end users.

The key element for success of this approach is fundamentally linked to a shared mindset of curiosity, systems thinking, and openness to feedback, coupled with robust expertise in each discipline.

The case studies illustrate the productive collaboration of the authors, underscoring the need for more evaluation with additional cases, as echoed by the literature (6).

An open question for ongoing practice is how to further enhance and focus team conversations through liberating structures as part of collaborative practice. With known processes and toolboxes, the emphasis should now shift towards enhancing self-organization and a collaborative mindset.

Further practical work is necessary as well as a concrete evaluation of the technical solutions of the case study, as far as this can be published.

References

1. Hammer, J., Kiesel, M.: Applying TRIZ to improve lean product lifecycle management processes. TRIZfest-2017 **81** (2017)
2. Lindemann, U.: "Creativity—a bottleneck in engineering design?" design research: The sociotechnical aspects of quality, creativity, and innovation, pp. 127–142. Springer International Publishing, Cham (2024)
3. Zierock, B., Jungblut, A., Senn, N.: "Chaotic customer centricity." International conference on human-computer interaction. Springer Nature Switzerland, Cham (2023)
4. Baschin, J., Inkermann, D., Vietor, T.: Agile process engineering to support collaborative design. Procedia CIRP **84**, 1035–1040 (2019)
5. Yang, C., Yuan, B., Junan, Y.E.: A user-centered development model for innovation design in automated nursing beds. J. Adv. Mech. Des. Syst. Manuf. **17**(5):JAMDSM0062-JAMDSM0062 (2023)
6. Chen, S., Kamarudin, K.M., Yan, S.: Analyzing the synergy between hci and triz in product innovation through a systematic review of the literature. Adv. Hum. Comput. Interact. 2021: 1–19 (2021)

7. Uhri, E., Isenhardt, I.: Function-oriented development of complex mechatronic products from an HTO perspective: A systematic literature review. Proceedings of the Design Society **3**, 2025–2034 (2023)
8. Van Pelt, A., Hey, J.: Using TRIZ and human-centered design for consumer product development. Procedia Engineering **9**: 688–693 (2011)
9. Lee, K.: "Innovative design thinking process with TRIZ." Automated Invention for Smart Industries: 18th International TRIZ Future Conference, TFC 2018, Strasbourg, France, October 29–31, 2018, Proceedings. Springer International Publishing (2018)
10. Sojka, V., Lepšík, P.: Use of Triz, and Triz with other tools for process improvement: A literature review. Emerg. Sci. J. **4**(5): 319–335 (2020)
11. Casner, D., et al.: "Agile 'TRIZ framework: towards the integration of TRIZ within the agile innovation methodology." Automated Invention for Smart Industries: 18th International TRIZ Future Conference, TFC 2018, Strasbourg, France, October 29–31, 2018, Proceedings. Springer International Publishing (2018)
12. Christian, P., Russo, D.: TRIZ industrial case studies: A critical survey. Procedia Cirp **39**: 51–56 (2016)
13. Koltze, K., Souchkov, V.: TRIZ-Anwendung in der Produkt-und Prozessentwicklung. Carl Hanser Verlag GmbH Co KG, Systematische Innovation (2017)
14. Nakagawa, T., et al.: World TRIZ Sites Project (WTSP) for Building and Maintaining Global Catalogs of TRIZ-related Web Sites. ETRIA TRIZ Future Conference 2018 (2018)
15. Fey, V., Rivin, E.: Innovation on demand: New product development using TRIZ. Cambridge University Press (2005)

Harmonizing Logic and Empathy by Integrating TRIZ and Design Thinking in Innovation

Jens Träger(✉)

Eppendorf SE, Hamburg, Deutschland
traeger.j@eppendorf.de

Abstract. This paper explores the integration of TRIZ (Theory of Inventive Problem Solving) and Design Thinking as a comprehensive approach to foster innovation and problem-solving. While TRIZ provides a systematic methodology for solving technical and inventive problems using inventive principles and patterns of innovation, Design Thinking offers a human-centered approach that emphasizes empathy, ideation, prototyping, and testing to meet user needs. By combining the structured, analytical approach of TRIZ with the empathetic, creative process of Design Thinking, this paper proposes a novel framework that leverages the strengths of both methodologies to tackle complex challenges. Through theoretical analysis and a practical application case, this study demonstrates how the integration of TRIZ and Design Thinking can lead to more innovative, user-centered solutions that are technically feasible and viable in the market. The aim is to provide practitioners and scholars with insights into how these methodologies can be synergistically applied to enhance innovation and user centricity.

Keywords: TRIZ · Design Thinking · User-Centric Solutions · Case Study

1 Introduction

Innovation stands as a critical determinant of competitive advantage and organizational success in today's rapidly changing business landscape. Traditional approaches to problem-solving and creativity often fall short when confronted with the complex, multifaceted challenges modern organizations face. This gap underscores the necessity for methodologies that not only foster creativity but also streamline problem-solving processes. The integration of TRIZ (Theory of Inventive Problem Solving) and Design Thinking emerges as a promising approach to offering a robust framework that combines logical rigor with empathetic creativity to enhance innovation efforts.

TRIZ, developed by *Genrich Altshuller* in the 1940s, is a systematic methodology designed to solve technical and inventive problems. It is based on the analysis of a vast array of patents and seeks to identify and apply inventive principles and patterns to overcome contradictions in technical scenarios [1, 2]. By providing a structured,

O. Mayer (Hrsg.): TRIZ-Anwendertag 2024,
https://doi.org/10.1007/978-3-662-70698-5_13

analytical approach, TRIZ allows practitioners to navigate complex problem spaces efficiently, focusing on the elimination of trade-offs and leveraging proven principles to generate innovative solutions.

In contrast, Design Thinking is a human-centered methodology that emphasizes understanding and empathizing with the user. It integrates stages such as ideation, prototyping, and testing, promoting a cycle of learning and adaptation that seeks to align solutions closely with user needs [3, 4]. This approach is especially effective in addressing vague or ill-defined problems where human factors play a significant role, making it invaluable in developing products and services that resonate deeply with users.

The potential for synergy between TRIZ and Design Thinking lies in their complementary strengths. While TRIZ offers a logical, principle-driven approach to innovation, Design Thinking provides a framework that is inherently flexible, adaptable, and user-centric. Integrating these methodologies can yield a comprehensive strategy that not only addresses the technical aspects of innovation but also ensures that solutions are desirable from a human perspective.

Though there are several methods than can be combined with TRIZ with sometimes unclear benefits and differentiation [5]. This paper gives a clear and easy to understand example in the form of a case study to show the link between Design Thinking and TRIZ in a form that can be used to facilitate other projects. The aim is to demonstrate how this integrated approach can tackle complex challenges effectively, leading to innovative, user-centered solutions that are technically feasible and commercially viable.

2 Literature Review on TRIZ and Design Thinking

TRIZ, a Russian acronym for Theory of Inventive Problem Solving, was developed by Soviet inventor *Genrich Altshuller* and co-workers starting in 1946. TRIZ is based on the premise that the evolution of systems is governed by certain patterns and principles which can be systematically applied to make inventive predictions and solve problems creatively [1]. Among the core methods of TRIZ are the 40 Inventive Principles and the Contradiction Matrix which helps innovators to identify and overcome specific contradictions in technological and engineering contexts.

Significant research has been conducted on TRIZ's application across various fields ranging from manufacturing to software development, demonstrating its versatility and effectiveness. Scholars like *Zlotin* and others [6, 7] have expanded on *Altshuller's* theories, applying them to complex systems and non-technical contexts, thus broadening the scope of TRIZ applications.

Design Thinking, popularized by *David Kelley*, founder of the design firm IDEO, and *Tim Brown*, emphasizes a human-centered approach to innovation. This approach is iterative and nonlinear, consisting of five phases: empathize, define, ideate, prototype, and test. These phases are designed to understand users deeply and to foster creativity in addressing their needs [8].

Design Thinking has been applied extensively in product and service design, with notable successes in sectors such as healthcare, education, and information technology. The methodology's strengths lie in its flexibility and focus on user engagement, which helps ensure that solutions are both innovative and aligned with user needs. Studies such as those by *Johansson-Sköldberg et al.* [9] delve into the theoretical underpinnings of Design Thinking, exploring its roots in design and its impact on practice.

While TRIZ and Design Thinking have been developed independently, there is growing interest in exploring how these methodologies can be combined. The integration aims to utilize the structured, analytical capabilities of TRIZ along with the empathetic, iterative processes of Design Thinking. Such a combination promises to enhance both the efficiency and the human-centeredness of the innovation process [10–12]. This approach has already been incorporated into textbooks on modern TRIZ techniques [13]. For example, the analytical tools of TRIZ can be used to inform the ideation phase of Design Thinking, potentially leading to solutions more tailored specifically to user needs.

3 Methodology and Procedure

Building on the complementary strengths of TRIZ and Design Thinking, a simple yet meaningful example will show how these methods interact with each other. This integrated framework is not only practical in its application but also significant in its theoretical implications. It bridges the gap between engineering-driven and human-centered approaches, potentially leading to a new hybrid innovation methodology.

As an example the Wallet Exercise, featured on programs at the d.school's Design Project Zero (DP0) [14, 15], serves as an immersive introduction to the principles of innovation and human-centered design. Usually, participants work in pairs within larger groups, engaging in a 90-min session complemented by preparatory steps. Each group is equipped with materials like worksheets, basic craft supplies such as markers, paper, and pipe cleaners, which are detailed on the d.school's website [15].

The exercise begins with empathetic interviewing. One participant assumes the role of an interviewer to delve into the experiences and needs of the other, who acts as a customer. The discussion centers around the customer's wallet, examining its contents and usability issues, and identifying any friction points. This phase is pivotal for building empathy and understanding the customer's daily interactions with their wallet. Following the interviews, participants articulate the latent needs and opportunities for innovation identified during their conversations. They frame these needs in a structured sentence that combines a specific user requirement with an emotional resonance and a key insight, like the need for security in tracking wallet contents to alleviate the anxiety of losing it. Participants then shift to generating concepts. They brainstorm several innovative solutions that might not even resemble traditional wallets but are designed to address the discovered needs. Next, the exercise moves into its most hands-on phase: prototyping. Using the provided materials, participants create basic prototypes of their ideas. The goal here is not to achieve perfection but to build something tangible enough to evaluate and discuss. Finally, participants use their storytelling skills to pitch their new wallet concepts. They explain how their designs meet the identified needs, aiming to persuade their partner or the larger group of their solution's viability. This activity emphasizes the journey over the destination. Engaging in the Wallet Exercise offers participants a concrete experience of the innovation process, from empathy and problem identification to prototyping and pitching. The debriefing part, where participants share and discuss their prototypes, underlines the importance of empathy, creativity, and the necessity of seeking feedback early and often in the design process.

To serve the purpose of this paper, a typical run through the wallet exercise has been compiled from typical outcomes of various participant groups. This filled-in wallet exercise, which is attached as supplementary material, shows the interaction of Design Thinking and TRIZ in the next section.

4 Case Study "Wallet Exercise"

Here is a practical example of how the results from the "wallet exercise" can be transformed into an input for TRIZ problem solving and how solutions beyond the output of the Design Thinking Process can be achieved. Summarizing the Design Thinking process, the following main problems, or tasks were identified:

1. **Bulk and Profile:** Users find that bulky wallets disrupt the silhouette of their attire, making them uncomfortable, especially when wearing fitted clothing. This also influences their choice of wardrobe, highlighting a need for a design that is both functional and stylish.
2. **Organization and Accessibility:** There is a strong need for wallets to be organized in a way that allows easy access to essentials like cards, cash, and coins without fumbling. Users prefer clearly defined compartments to avoid clutter.
3. **Comfort and Weight:** A lightweight wallet is crucial for daily comfort and ease of carrying, with an emphasis on not altering the wearability of various attire.
4. **Aesthetics and Material Quality:** Users associate the feel of the wallet's material with its overall quality. There is also a significant preference for wallets that maintain the aesthetic appeal of their clothing.

The wallet challenge also outlines several solutions to address the issues identified during the Design Thinking process:

1. **Modular Wallet Design:** Introducing a wallet with detachable sections allows users to customize their wallet based on their needs for different occasions. This design offers flexibility without compromising the slim profile of the wallet.
2. **Simplified Interior Layout:** Reducing the number of compartments and designing them to be more accessible can improve user experience. Essential compartments are ergonomically arranged to offer ease of use without sacrificing functionality.
3. **Material Upgrade:** Using high-quality, durable materials ensures the wallet withstands daily wear and tear while maintaining a luxurious appearance.
4. **Customizable Features:** Including options like RFID-blocking capabilities, integration of a smart tracker for those prone to misplace their belongings, and eco-friendly materials to appeal to environmentally conscious consumers.
5. **Additional Innovations:** Proposals such as a sleek, detachable LED light panel for ease of use in low light conditions further enhance the wallet's functionality and appeal.

These solutions collectively aim to create a wallet that is practical, stylish, and adaptable to the varying needs of users, reflecting a thorough understanding of the target market's preferences and requirements.

Fig. 1: Drawing of the final result prototype of the Design Thinking

Design Thinking has shown effectiveness in identifying and creating solutions that might not typically emerge from TRIZ, particularly in areas of problem definition and resolution. Throughout the process, it became evident how crucial aesthetics are to the user persona, *Sophia*. Notably, the final outcome at the end of the process was the integration of a wallet and a cellphone cover into a decorative accessory using materials like gold, (synthetic) diamonds, and pearls, designed to be visibly worn as jewelry (Fig. 1). This approach does not address any contradictions as TRIZ might, but rather transforms a deficiency into a unique feature. Nevertheless, Design Thinking can sometimes fall short in generating highly innovative and inventive solutions. Most of the other solutions do not solve the problem identified completely, as can be seen from the drawings. The proposed solution does not really overcome issues like bulkiness.

However, Design Thinking readily brings up points for improvement. This becomes obvious, when we compare the function analysis like it would have been done by a person trained in TRIZ (Table 1) compared to the function analysis made through the evaluation of the Design Thinking process (Table 2). The main function of a wallet is holding personal items, which are typically cards, cash, and coins. From a technical perspective the interactions between the components of the wallet and super-system components are useful interactions as all proposed wallets will hold the user's personal items in a reasonable manner. From the empathizing with the user during the Design Thinking process, we learn that those useful functions have certain drawbacks. As users desired a slim wallet to avoid disrupting the silhouette of their outfits, in a TRIZ context, the function of "holding" the contents can be modeled as excessive, as it contributes too much bulk to the wallet's construction. The volume and thickness of the wallet, making it bulgy to wear and the resulting disturbance of the user can easily model as harmful functions. The desire of the users to have an RFID (i.e. radio frequency) blocking wallet or a certain haptic of the outer material are also not necessarily obvious. In the TRIZ way of doing a function analysis, both would be useful but insufficient functions if the wallet would not have the required features. The prerequisite for a correct user centric modelling of the function analysis is the availability of that information.

Every interaction that is useful but insufficient or excessive as well as harmful interactions from a function analysis reveals problems that provide opportunities for improvement by solving them with methods from TRIZ. The Design Thinking enhanced function analysis not only reveals more of such opportunities, it also ensures that the right problems are addressed, as these are derived directly from empathizing with the user.

Table 1: Function analysis based on technical factuality.

Subject	Action	Object	Type of Interaction
Wallet	**Holds**	**Personal Items**	**Useful, normal**
Card Slots	Hold	Cards	Useful, normal
Bill compartment	Holds	Cash	Useful, normal
Coin Pocket	Holds	Coins	Useful, normal

Table 2: Function analysis based on the findings from the Design Thinking process.

Subject	Action	Object	Type of Interaction
Wallet	**Holds**	**Personal Items**	**Useful, insufficient**
Card Slots	Holds	Cards	Useful, excessive
Bill compartment	Holds	Cash	Useful, excessive
Coin Pocket	Holds	Coins	Useful, excessive
Wallet compartments	Organize	Contents	Useful, insufficient
Outer Material	Blocks	Radio signals	Useful, insufficient
Outer Material	Delights	User	Useful, insufficient
Wallet	Bulges	Apparel	Harmful
Wallet	disturbs	User	Harmful

The main issue identified is transformed into a physical contradiction for resolution using the TRIZ method, which follows a structured and staged approach for the resolution of physical contradictions and applying 40 inventive principles [16, 17]. This approach tends to yield more and better solutions than the Design Thinking Process.

Physical contradiction:
The wallet should have a large volume in order to carry all the personal items and the wallet should have a small volume in order to keep the silhouette of the attire undisrupted.

1. Separation in Space
Inventive Principle #1 – Segmentation: The wallet is segmented into expandable or detachable compartments that can be modified according to the user's needs. We can design a wallet with adjustable compartments that expand, and contract based on the user's immediate needs. The wallet could employ materials that are inherently flexible and expandable while still retaining a flat, compact shape when not fully utilized. Technologies like soft, stretchable fabrics or elastomers that are sturdy yet flexible could be employed. Alternatively, an accordion-style folding mechanism could be employed for the same purpose. When fewer items are needed, these sections can remain flat. Alternatively, we could consider a modular design where different sections (for cards, cash, coins, etc.) can be detached or attached as needed, allowing users to customize the size and content of their wallet according to the occasion.

The wallet could be split into separate wallets, e.g. one for cards and on for cash that could be worn on different parts of the body, e.g. different pockets to avoid excessive bulging and disturbing the silhouette of the attire.

Inventive Principle #3 – Local Quality: Different parts of the wallet have different properties or functions, such as areas that expand more than others. The wallet is made of elastomeric or flexible materials.

2. Separation in Time
Inventive Principle #10 – Prior Action: Preparing the wallet in different configurations ahead of time based on anticipated needs. This principle could be used by encouraging the user to carry different items at different times depending on specific needs. For example, during work hours, a wallet could be configured to carry more business-

related items like business cards and a pen, while for a night out, it could be slimmed down to just include a credit card and some cash. This not only reduces the wallet's size at any given time but also ensures it contains only the essentials required for specific situations.

Inventive Principle #15 – Dynamization: Making parts of the wallet adjustable so they can change or adapt as needed over time.

3. Separation in Relation
Inventive Principle #2 – Taking out: Separating elements that serve different functions (professional vs. personal items) into different parts of the wallet. Applying this principle, the wallet's components could be designed to serve different purposes for various aspects of the user's life. For example, one part of the wallet could be dedicated to professional life (holding business cards and RFID access cards), while another could be for personal use (cash, credit cards, and personal identification). This increases the level of organization in the wallet, as each section could be used independently, depending on the context.

4. Separation on Condition
Inventive Principle #15 – Dynamization: The wallet adapts its structure or capacity based on external conditions or usage, similar to the solution in step 1 (separation in space).

5. Satisfaction
Inventive Principle #30 – Flexible Shells and Thin Films: To fulfill both requirements simultaneously, the wallet could be constructed from flexible materials like elastomers. This results in a solution like in step 1 (separation in space).

6. Circumvention
Inventive Principle #28 – Replace Mechanical System: The need for a traditional and potentially disturbing wallet could be eliminated entirely by embracing digital wallet technologies. This would involve using mobile devices to store digital versions of cards, identification, and tickets, removing the reliance on physical wallets. This approach aims to maximize benefits with minimal drawbacks, transitioning toward a more virtual wallet experience. This solution corresponds to the general principle of increasing ideality in that the function of the wallet becomes field-like through the transition to a digital supersystem component (i.e. cell phone).

It is obvious that the separation principles do not all lead to independent solutions. However, this example demonstrates the benefit of solving physical contradictions thoroughly by going through all solution principles, as the circumvention approach offers a unique solution: replacing the mechanical wallet with a "field-like" digital version, and transfer to a super system component such as a cell phone app. Finaly, we have three independent technical solutions:

1. The wallet is made flexible, using elastomeric material or an accordion-like design.
2. The wallet consists of detachable sections that can be combined or removed as needed. Interestingly, the separation in relation approach surfaces the added benefit of enhancing the wallet's organization.

3. The wallet is replaced by digitalization, using a cell phone to store digital versions of cards, identification, tickets, etc.

In summary, the TRIZ methodology has identified valuable solutions that have not been found through Design Thinking alone.

5 Discussion

The integration of TRIZ and Design Thinking as outlined in the proposed framework has demonstrated significant potential in the case study presented. The case study highlights that the integration of TRIZ and Design Thinking can substantially enhance the quality of the solutions developed. By systematically addressing both the technical contradictions and the user-centered design aspects, the solutions not only meet the technical requirements but also resonate deeply with user needs. Moreover, the structured approach provided by TRIZ can significantly accelerate the innovation process by offering clear pathways to overcome specific barriers, which, when combined with the iterative cycles of Design Thinking, ensures that solutions can be rapidly prototyped and refined. Moreover, the example shows, that TRIZ delivers not only more but more inventive solutions than the Desing Thinking process alone. For example, the transition of the wallet to digital world as an app on a cell phone is much more likely to be found by the structured approaches from TRIZ than with the basically non-directional brainstorming during the Design Thinking approach.

The empathetic approach of Design Thinking allowed for deeper insights into user needs and behaviors, which were effectively incorporated into the solution design. This alignment with user expectations not only facilitated a smoother adoption process but also enhanced the overall satisfaction with the end products. The framework's emphasis on iterative testing and feedback integration ensures that the final products are well-tuned to user needs, potentially increasing the market success of the innovations.

Practically, this example offers a structured yet flexible approach that can be customized to fit various organizational needs and innovation goals. It encourages a creative yet disciplined approach to innovation, which is crucial for maintaining competitive advantage in fast-paced markets. By engaging with both TRIZ and Design Thinking, teams develop a more holistic view of problem-solving that combines technical proficiency with user-centric insights. This dual focus can cultivate a more adaptable and innovative organizational culture, and therefore contributes to organizational learning.

Despite the apparent benefits, integrating TRIZ and Design Thinking is not without challenges. One major hurdle is the potential cultural clash between the systematic, engineering-focused TRIZ users and the more intuitive, design-oriented thinkers. Bridging this gap requires not only training and adaptation but also a shift in mindset to appreciate and leverage the strengths of both methodologies. Organizations must be prepared to invest in skills development and process management to fully realize the benefits of this integrated approach.

6 Conclusion and Outlook

The exemplary case study shown here provides a comprehensive integration of TRIZ and Design Thinking within the context of product innovation. As industries continue to evolve and new challenges emerge, the flexibility and robustness of this integrated approach can be critical in shaping future innovation.

Both TRIZ and Design Thinking offer unique advantages, their integration can potentially yield a more holistic approach to innovation. This synergy can help bridge the gap between technical feasibility and user desirability, thereby creating solutions that are not only inventive but also deeply resonant with user needs. Among the synergistic benefits are Enhanced Innovation, Balanced Solutions, and Iterative Improvement. By leveraging TRIZ's systematic approach and Design Thinking's user-centricity, the framework fosters more creative, practical, and sustainable innovations. Solutions will not only be technically robust but also deeply resonant with user needs and expectations. The cyclical nature of the Design Thinking framework promotes continuous improvement and adaptation, crucial for keeping up with dynamic market and technology environments.

Future research could explore the scalability of this framework across different industries for approaching and tackling complex problems to understand broader applicability and potential limitations. Comparative studies could be conducted to evaluate the effectiveness of the integrated approach against traditional single-methodology approaches.

Acknowledgement. The author extends his deepest gratitude to Lidiya Mestechkina for her invaluable assistance and expertise in the Design Thinking wallet challenge.

References

1. Altshuller, G.: Creativity as an Exact Science. Gordon & Breach Science Publishing House, New York (1984)
2. Petrov, V.: TRIZ Basics. Independently published (2020)
3. Lee, J.H., Ostwald, M.J., Gu, N.: Design Thinking: Creativity, Collaboration and Culture. Springer International Publishing, Cham (2020). https://doi.org/10.1007/978-3-030-56558-9
4. Tomlinson, M.D.: The Impact of Design Thinking on Driving Innovation Within Large Businesses (2018). https://doi.org/10.13140/RG.2.2.24520.62726
5. Mann, D.: Design Thinking – Philosophy, Method, Tool. Triz J. https://the-trizjournal.com/design-thinking-philosophy-method-tool/ (2018)
6. Zlotin, B., Zusman, A.: Managing innovation knowledge. TRIZ Prog. Ideation Int. Inc. 123–140 (1999)
7. Smith, L.R., Zlotin, B., Zusman, A.: The dilemma of improving quality in new product development. Altshuller Inst. Conf. TRIZCON99 7–8 (1999)
8. Brown, T.: Design thinking. Harv. Bus. Rev. **86**, 84 (2008)
9. Johansson-Sköldberg, U., Woodilla, J., Çetinkaya, M.: Design thinking: past, present and possible futures. Creat. Innov. Manag. **22**, 121–146 (2013)

10. Lee, K.: Innovative Design Thinking Process with TRIZ. In: Cavallucci, D., De Guio, R. Koziołek, S. (eds.) Automated Invention for Smart Industries, vol. 541, pp. 241–252. Springer International Publishing, Cham (2018)
11. Ocampo, J.U., Kaminski, P.C.: Proposal of an FFE model with a high degree of innovation integrating TRIZ and design thinking methodologies, specific for the personal health equipment sector. Cogent Eng. **10**, 2208943 (2023)
12. Da Silva, R.H., Kaminski, P.C., Armellini, F.: Improving new product development innovation effectiveness by using problem solving tools during the conceptual development phase: integrating design thinking and TRIZ. Creat. Innov. Manag. **29**, 685–700 (2020)
13. Orloff, M.A.: ABC-TRIZ. Springer International Publishing, Cham (2017). https://doi.org/10.1007/978-3-319-29436-0
14. Build Your Creative Confidence: The Wallet Exercise. https://www.ideo.com/journal/build-your-creative-confidence-the-wallet-exercise. Accessed 26 Apr 2024
15. Translated Resources. *Stanford d.school.* https://dschool.stanford.edu/resources/the-gift-giving-project. Accessed 26 Apr 2024
16. Gadd, K.: TRIZ for Engineers: Enabling Inventive Problem Solving. Wiley, Chichester (2011)
17. Mann, D.: Matrix 2010: Re-Update Der TRIZ Widerspruchsmatrix. C4pi – Center for Product-Innovation, Wennigsen/Deister (2013)

Building a Bridge between Design for Six Sigma and TRIZ

An Example from the Concept Phase of a Development Project

Achim Schmidt[1](✉) und Jürgen Moosburger[2]

[1] Innovigate GmbH, Regenstauf, Deutschland
achim.schmidt@innovigation.com
[2] ams-OSRAM International GmbH, Regensburg, Deutschland
juergen.moosburger@ams-osram.com

Abstract. Design for Six Sigma (DfSS) is an important and proven methodological approach for the development of robust products and is used at both system and subsystem or component level. The core idea of Design for Six Sigma is to develop products from the very beginning in such a way that they are later robust in series production and achieve Six Sigma quality. A major focus is also placed on the definition (elicitation), fulfillment and verification of customer and stakeholder requirements. In the early phase of concept development, DfSS methods such as the Boundary Diagram support the creation of the system architecture. This paper uses an example to show how the quality of the Boundary Diagram can be significantly improved with the help of TRIZ methods and how risks and failures can be identified even before real prototypes are created. This example is intended to illustrate how a bridge can be built between the two methodological approaches and the benefits that can arise.

Keywords: TRIZ · DfSS · Design for Six Sigma · System architecture · Robust design · Robust products · Function Analysis · Boundary Diagram · Anticipatory Failure Determination

Design for Six Sigma (DfSS) is "one of the most common methodological roadmaps for developing efficient and robust products and processes in a short time" [1].

TRIZ—the theory of inventive problem solving—comprises many methods and tools for creating inventive solutions to problems, based on the research and findings of Genrich Altshuller (see e.g. [2, 3]). The methodology is constantly being further developed and is used in many companies, particularly to increase innovation and solve technical problems.

Design for Six Sigma contains many proven structuring and statistical methods, but the proportion of innovation methods is rather small. TRIZ methods can close this gap in an excellent way.

O. Mayer (Hrsg.): TRIZ-Anwendertag 2024,
https://doi.org/10.1007/978-3-662-70698-5_14

This article uses an example to describe how the integration of TRIZ methods can be used to find new ideas in the early phase of designing the architecture of a mechatronic system with the help of TRIZ methods. In this example, safety gaps are even identified and closed (Fig. 1).

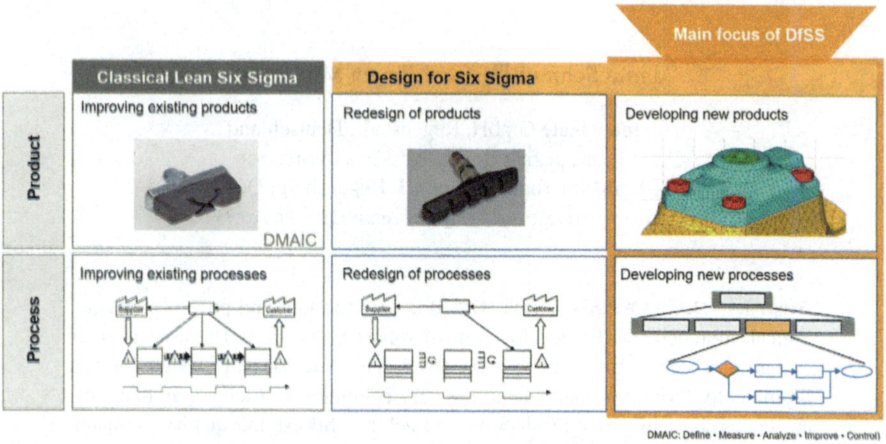

Fig. 1 Application of the Lean Six Sigma and DfSS approaches in comparison

The common application of TRIZ and DfSS methods has already proven itself in practice in many projects. For reasons of confidentiality, this publication does not describe a real project, but uses a case study to illustrate the methods used in the DfSS Green and Black Belt training courses. This case study is the result of a cooperation

Fig. 2 Electronic tailgate actuator with remote key

between the consulting companies Innovigate GmbH and Process Fellows GmbH. This case study is about the actuator unit of a tailgate drive for cars. It is a mechatronic system comprising hardware (electronics and mechanics) and software (Figs. 2 and 3).

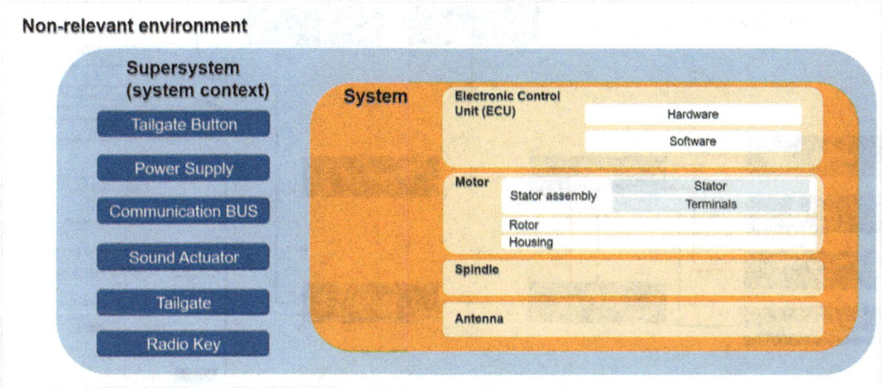

Fig. 3 Rough draft for the system architecture

In the early phases of a DFSS project, a Boundary Diagram is used to define the boundaries for the system. In this diagram, the various system elements, interfaces are defined and the supersystem elements that are relevant to the system are shown. Differently marked arrows symbolize connections between the system and supersystem elements, such as physical connections, transfer of electrical energy, transfer of mechanical energy, transfer of material, transfer of information. (see e.g. [4, 5]).

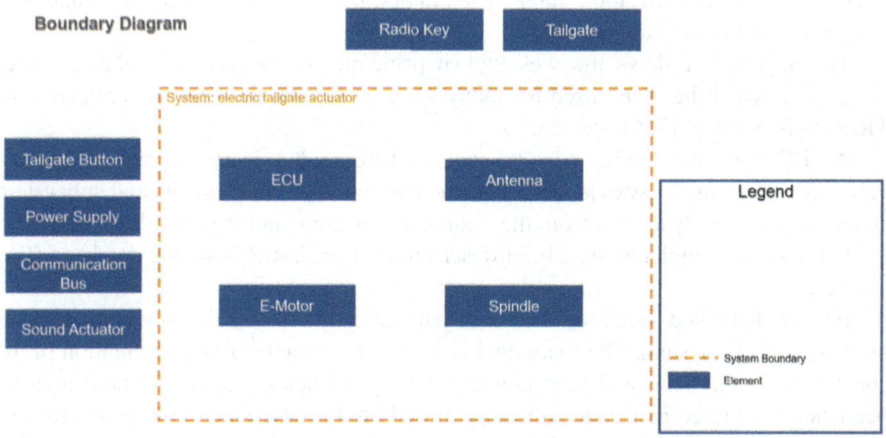

Fig. 4 First draft of the Boundary Diagram with system and supersystem elements

Besides the requirements analysis tools, the Boundary Diagram is one of the most common and important tools in DFSS projects, as it forms the basis for the system architecture. During the development of the Boundary Diagram, many decisions are made regarding the system design (Fig. 4 and 5).

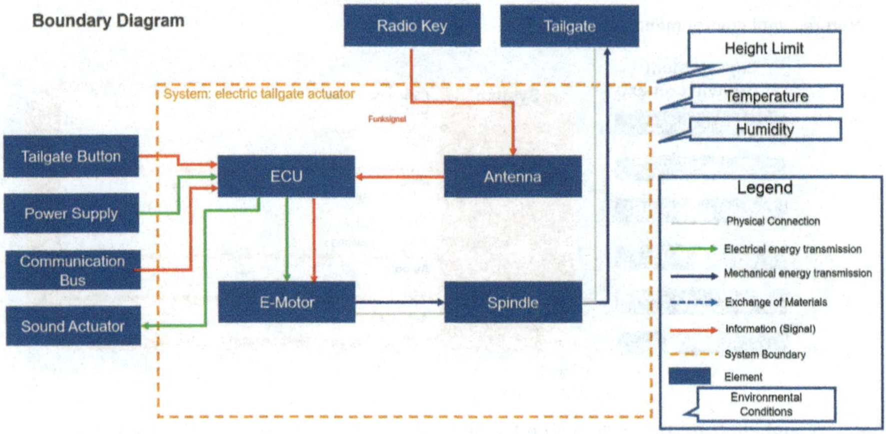

Fig. 5 Boundary Diagram for the electronic tailgate including system interactions

However, the diagram is only a descriptive tool that documents the structure and dependencies of the system components. The Boundary Diagram is used at an early stage of the development project to identify the system elements, interfaces, and connections—but at the same time all this information is defined immediately. Therefore, it is not very likely that the number of system elements defined in the Boundary Diagram can be changed later.

The TRIZ Function Analysis is a very powerful tool for improving the system. This tool and other TRIZ tools can be used to optimize the boundary diagram and thus to optimize the system architecture.

The approach follows the well-known principle of abstraction—solution—specialization, which has been used for many years to explain the general application of TRIZ tools (see e.g. [5, 6] and Fig. 6).

A TRIZ Function Analysis can be generated from a boundary diagram with relatively little effort, as the system elements, such as supersystem elements and subsystem elements, are already known from the system architecture and the boundary diagram.

For sake of completeness, a brief description of the TRIZ Function Analysis (FA) is given:

FA is a dedicated TRIZ tool used to systematically analyze the functions of components within a system. The main goal is to obtain a (graphical) representation of the constituent components and their mutual interdependencies. Moreover, these interdependencies are rated in a systematic way. Based on this assessment, the points for improvement can be highlighted. Goals can be e.g. increase of functionality, prevention of harmful or insufficient effects, or cost reduction.

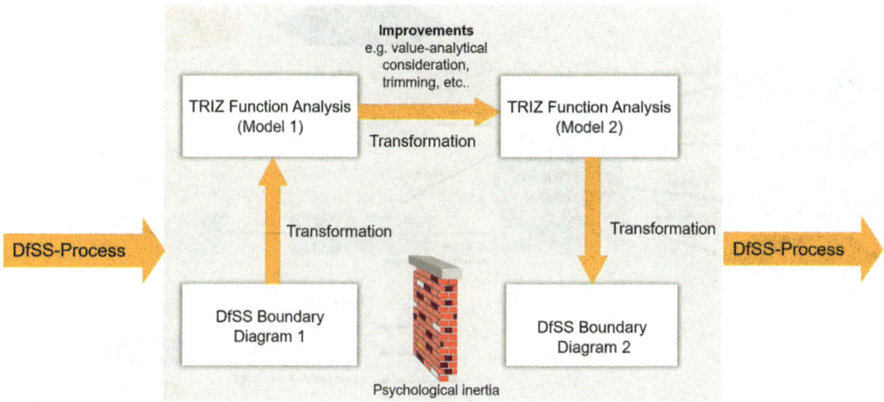

Fig. 6 Creating a better boundary diagram using the TRIZ approach, see also [5]

The 3 main steps of a basic FA are:

Component Analysis:

In this step, the system model is decomposed into its main components or parts. The level of decomposition must be consistent (all listed components on the same system hierarchy) and depends on the problem to be solved.

Interaction Analysis:

In this step, the focus is on mapping the mutual interactions between the components in tabular form without further specification. This is done in the next step.

Functional Modeling:

Functional modeling aims to capture the essential functions performed by each component. As an interaction is verbally represented by 'subject—verb—object', a function is closely related to an active verb. This is important to distinguish between function 'carrier' and 'object' of the function.

After having done the basic FA, a complete functional model is built up (Fig. 6, model 1) and the weak points are identified and development task can be formulated. The advanced FA tools dealing with that are the Trimming and the Cost-Value Analysis. Solving these tasks can lead to an (virtual) improved technical system (Fig. 6, model 2). This is now applied to the given example of an actuator unit of a tailgate drive for cars.

In this example, Anticipatory Failure Detection is used to identify malfunctions and safety risks as early as the concept phase. Keyless entry systems pose the threat of car break-ins or even theft. The task now is to look for ways to prevent hackers from gaining unauthorized access.

The TRIZ function analysis in Fig. 7 shows the undesired function: a hacker uses a modified radio key to send an unauthorized signal to the system, which is received via the antenna. This signal causes the tailgate to open without authorization.

In fact, it is usually more difficult for development teams to think about risks and undesired functions than to develop desired functions. To overcome this psychological inertia, the TRIZ tool of predictive failure detection is used in this case study.

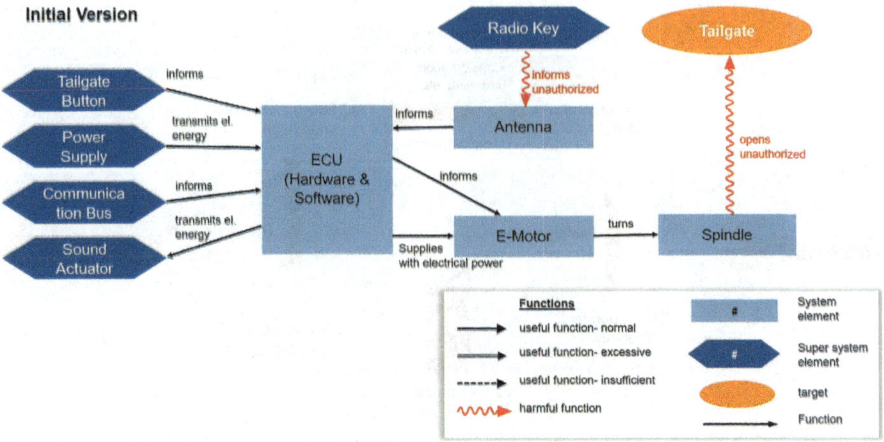

Fig. 7 TRIZ functional analysis for the ETG system exposed to a hacker attack

The TRIZ-Anticipatory Failure Determination is a systematic problem-solving method derived from the TRIZ principles, which is used to predict and analyze potential failures in a system or process. The benefits of this tool include enhancing reliability and safety by proactively identifying weaknesses, thus allowing for improvements and preventative measures before failures occur (for detailed information see [7]).

With this method, the undesired function is converted into a desired function. In this case, the question is not how unauthorized access to the opening system can be prevented, but the opposite question is asked: how can the system be opened by unauthorized access? With Anticipatory Failure Determination, this question is reinforced in the next step and is therefore: how can a hacker reliably open the tailgate of the car using a digital attack (Fig. 8)?

This type of unauthorized access can be achieved by tampering. Tampering refers to the unauthorized alteration or manipulation of a device or system, often with the intent to bypass security measures or cause malfunction. In the context of a remote key-

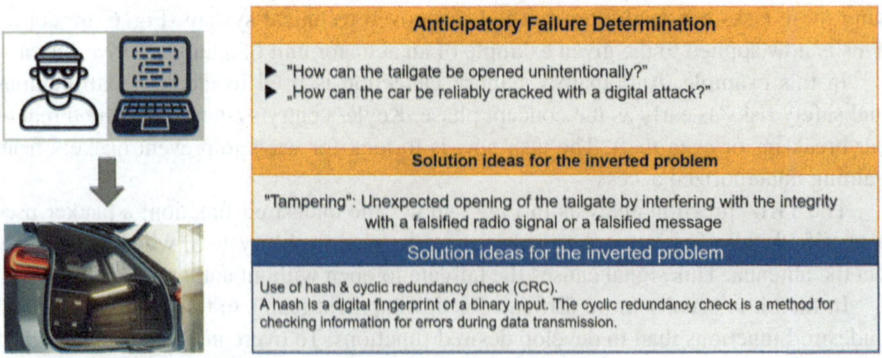

Fig. 8 Anticipatory Failure Determination for the electrical tailgate (ETG)

less entry system for a car, tampering can mean intercepting or duplicating the signal of the key fob to gain unauthorized access to the vehicle.

The next step in Anticipatory Failure Determination is to convert the solutions found for the undesired function into measures to prevent the undesired state and into the desired functions.

To protect against tampering, keyless systems can be secured e.g. with advanced encryption methods and frequency hopping to prevent signal interception.

In this example a keyless entry system using hash functions and cyclic redundancy check (CRC) is implemented. It enhances security and integrity of the signal transmission between the key fob and the vehicle. The system uses hash functions to generate a unique, fixed-size cryptographic representation of the data transmitted, ensuring that any tampering or alteration is detectable. CRC is added to this process to check for any accidental changes to the raw data, such as those caused by noise or transmission errors, ensuring the reliability of the data when it reaches the vehicle's receiver (Fig. 9).

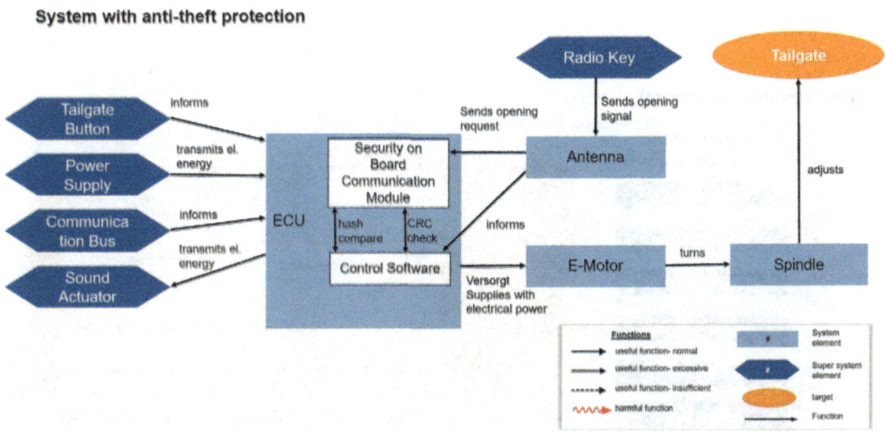

Fig. 9 Optimized TRIZ functional analysis

The use of the hash and cyclic redundancy check procedure has an effect on the boundary diagram and on the system architecture shown in Figs. 10 and 11, because the architecture of the electronic control unit (ECU) is detailed. It now contains a control module and a security module and has corresponding communication paths.

Conclusion

The integration of the TRIZ functional analysis into the Boundary Diagram enables a more focused and structured approach within a DfSS based product development. This enables prototypes with a high degree of maturity in the early phases of product development.

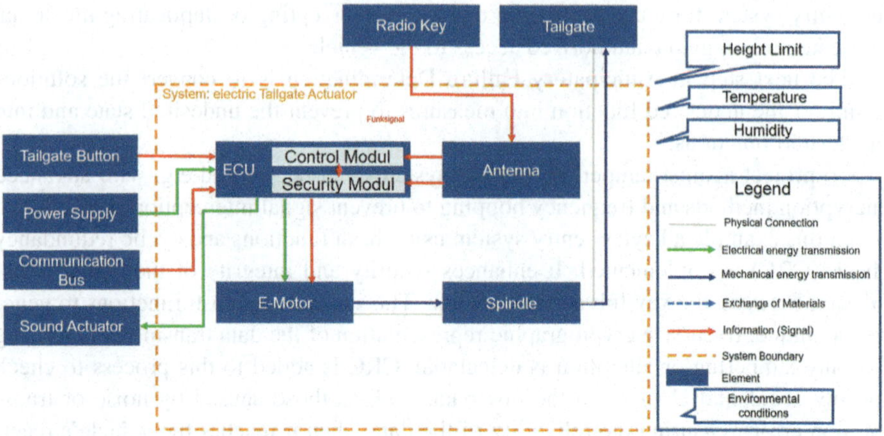

Fig. 10 Boundary Diagram including ECU control and security modules

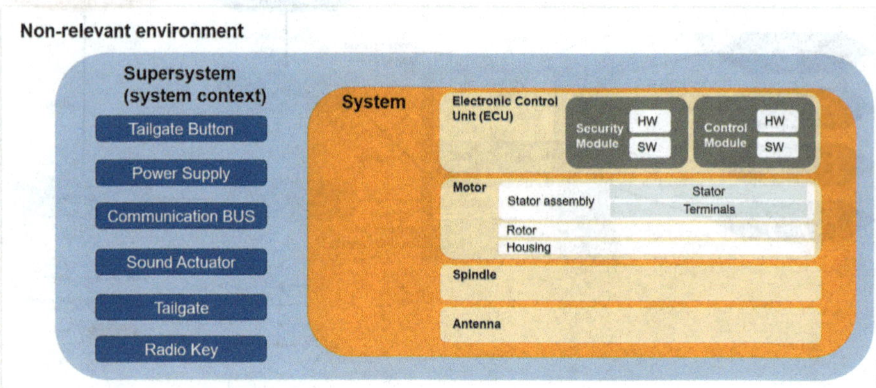

Fig. 11 System architecture including ECU control and security modules

In addition, TRIZ methods are used to proactively identify technical risks and weaknesses, find opportunities for innovation and solve technical challenges.

This was demonstrated using the example of an electric tailgate drive, in which a security gap was already discovered and closed during the concept phase (Fig. 12).

Acknowledgements. The authors are grateful for the support of the management consultancy Process Fellows in the joint development of the case study "Electronic Tailgate."

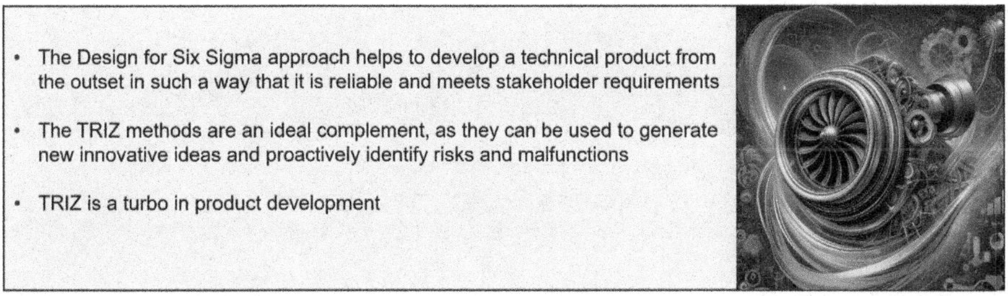

Fig. 12 Key point summary

References

1. Cudney E.A., Agustiady T.K.: Design for Six Sigma: A Practical Approach Through Innovation. CRC Press. ISBN 9781498742559 (2016)
2. Altshuller G.S.: “Creativity as an exact science”. Moscow, Sovietskoe ratio, Translated by Williams A. 1984, Crc Pr Inc, ISBN 978-0677212302 (1979)
3. Altshuller G.S.: And suddenly the Inventor appeared. Technical Innovation Center Inc., Translated by Shulyak L., ISBN 0-9640740-2-8 (1996)
4. Konert, T., Schmidt, A.: Design for six sigma umsetzen. Hanser, München (2010)
5. Schmidt, A., Thurnes C.M., Näther, S.: Hybridization of DfSS- and TRIZ-Modeling Methods within an innovation oriented DfSS framework, TRIZfest 2017, Krakow, Poland (2017)
6. Kaplan S.: An Introduction to TRIZ – The Russian Theory of Inventive Problem Solving. Southfiled. ISBN 1-928747-0-00 (1996)
7. Svetlana Visnepolschi, Der innovative Weg zu Null Fehler – Aktuelle TRIZ-Methoden der Antizipierenden Fehler Erkennung AFE, Synnovating, 978-3-9815493-1-7

How to Tailor Generative AI Prompts for the Application of Inventive Principles

A Part of the Collaborative and Creative TRIZ Open-Prompts Project (ccTOPP)

Jürgen Moosburger[1(✉)] und Robert Adunka[2]

[1] ams-OSRAM International GmbH, Regensburg, Deutschland
juergen.moosburger@ams-osram.com
[2] TRIZ Consulting Group GmbH, Sulzbach-Rosenberg, Deutschland
adunka@triz-consulting.de

Abstract. Within the collaborative and creative TRIZ Open-Prompts Project – ccTOPP – the capabilities of generative AI for TRIZ based problem solving are investigated. The aim is to establish specific customized prompts for the majority of the TRIZ tools, that will be then freely available. In this article, we present the current status of our investigations on how to tailor generative AI prompts for the application of inventive principles. It shows that one has to carefully compose the priming prompt in order to increase the quality of prompting results. The project links TRIZ experts all over the world, thus contributes to the global TRIZ community. The authors are active members of the European TRIZ-Campus (ETC).

Keywords: TRIZ · Inventive Principle · Sub-Principle · Generative artificial Intelligence (genAI) · Prompt · Prompting techniques · Priming · Large language model (LLM) · Zero shot · One shot · Few shot · Chain-of-thought · Least-to-most · Tree of thought · ChatGPT · Microsoft Co-Pilot

1 Introduction into TRIZ & Generative AI

TRIZ (Theory of Inventive Problem Solving) is a systematic methodology for problem solving and innovation. It provides a structured approach to identify and to overcome technical challenges by using dedicated analysis and ideation tools derived from an extensive patent analysis. On the other hand, generative AI (genAI), a subset of artificial intelligence (AI), focuses on creating new content based on large language models. It generates new data instances by a word-to-word probability rating that is based on the given input/training data. In a first view, the combination of TRIZ and

O. Mayer (Hrsg.): TRIZ-Anwendertag 2024,
https://doi.org/10.1007/978-3-662-70698-5_15

genAI might be able to revolutionize the process of problem-solving and innovation by automating the generation of creative and effective solutions. Having a closer look then reveals the challenges of that approach. Ideation is a complex creative, cognitive and associative process. So how can a word-by-word probability rating based on a high amount of input data complement or even mimic the mental processes of a genius expert in a specific field of investigation? Therefore, it is firstly essential to have a profound understanding of architecture of genAI and LLM, and the way of utilizing them. Secondly, interacting and utilizing genAI for TRIZ based problem solving has to be investigated so that practical knowledge can derived and summarized. This is the goal of the ccTOPP project.

2 Fundamentals & Definitions of genAI prompting [1, 2]

Generative AI prompting is an approach to guide generative AI models and control their output to produce desired high-quality results.

In the following, the key definitions are given, that are fundamental to get deeper into the topic.

- **Definition:**
  Large Language Models (LLM) are machine learning models that are particularly powerful in natural language processing (understanding and generating) due to their size and complexity.
- **Tasks:** Understanding and generating human-like language.
- **Applications:** text generation, machine translation, chatbots, text analysis.
- **Challenges:** Training and operation require significant computational resources, ethical concerns about potential for abuse.
- **LLM Settings:**
  - Temperature (0 = not creative, 1.5 = creative)
  - Top P (randomness of the output: 0.9 = 90% – tokens that add up to 90% are taken into account)
  - Maximum Length (the maximum number of tokens the AI is allowed to generate)
- **Tokens:** In ChatGPT, **tokens** are the smallest units in processing. Usually, a token is equal to one word. The AI generates text by predicting the next one based on previous tokens. The AI considers all tokens at the same time and does not read from left to right.
- **Context Length:** The **Context Length** is the number of previous tokens that the model is looking at. In GPT-3.5 Turbo, the maximum context length is 4096 tokens.
  At this boundary, older parts of the context are removed to make room for new tokens.

Beside expressing your query as clearly as possible it is essential to understand how genAI models work, what input they need and which limitations are to be considered.

As Tanasak Pheunghua (also ccTOPP project member) states in his article "TRIZ and Generative AI – Example of Prompt" [3]: "It is important to recognize that Generative AI models, …, are not truly "thinking" in the same way humans do. They are essentially complex mathematical functions that process and recombine large datasets to generate new content ….These models do not possess genuine comprehension or reasoning capabilities like the human brain. Therefore, their outputs should be viewed as supportive inputs to the human creative and problem-solving process, rather than as a replacement for human expertise and decision-making."

So experimentation is key to come up with best practices. The goal is to establish guidelines of how to treat, respectively 'communicate' with LLM in a way to get as accurate and specific answers as possible.

LLMs that are under investigation with ccTOPP are:

OpenAI chatGPT: https://openai.com/
Google Gemini: https://gemini.google.com/app
Microsoft Bing/Copilot (chatGPT-4): https://www.bing.com/

Brief Explanation of the most important prompting techniques

In the following, the most prominent prompting techniques are briefly outlined [4]. Having them in mind is key to understand the current practical investigations within the ccTOPP project as explained in the last chapter of this article.

1. **Zero-Shot Prompting**: Providing the AI with a task or instruction without any additional context or examples.
2. **One-Shot/Few-Shot Prompting**: Providing the AI with one or a small number of relevant examples or demonstrations to guide its response.
3. **Role Prompting**: Specifying a persona or perspective for the AI to adopt, such as an expert, artist, or historical figure.
4. **Chain-of-Thought (CoT) Prompting**: Instructing the AI to provide a step-by-step reasoning process to arrive at the final answer, rather than just the result.
5. **Zero-Shot Chain-of-Thought**: Combining zero-shot prompting with the chain-of-thought approach, where the AI is asked to reason through a task without any additional context.
6. **Tree of Thought**: An extension of Chain-of-Thought prompting, where the AI is asked to explore multiple possible lines of reasoning or solutions in a branching, tree-like structure.
7. **Least to Most Prompting:** The problem to be solved is broken down into a series of sub-problems that build on each other. In a second step, these sub-problems are solved one by one. Unlike the Chain of Thought, the solution to the previous sub-problems is fed into the prompt that attempts to solve the next problem.
8. **Other Techniques:** There are various other prompt engineering methods, such as using specific language patterns, incorporating visual or multimodal inputs, and leveraging knowledge bases or external information sources (Fig. 1).

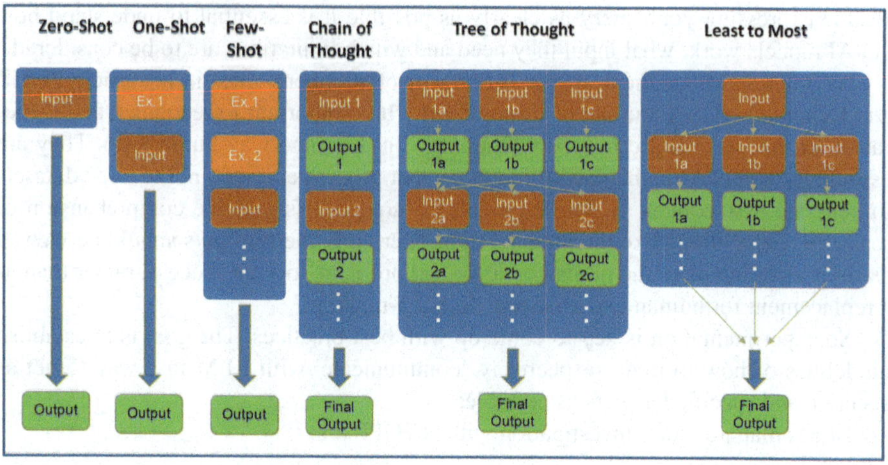

Fig. 1 Overview over different types of prompting techniques

Challenges, Pitfalls & Risks

Generative AI prompting, while powerful, does come with certain Risks. The most important are listed below:

1. **Accuracy**: Generative AI models like GPT-3 rely on large language models trained on massive data. While they can generate plausible-sounding responses, they may sometimes produce inaccurate or false information1. They may also generalize over specificity, which can lead to inaccuracies for nuanced or out-of-sample queries [5].
   - **Hallucinations:** AI can generate false information when asked for information it doesn't know. Instead of simply explaining "I don't know," the AI often simply generates text that sounds perfectly plausible, but isn't true at all. E.g. "Explain the 9-field thinking at TRIZ".
   - **Mathematics:** Problems from mathematics often cause problems for AI. False results can occur even with simple multiplication tasks. You're trained on a lot of text, but math requires a different approach.
2. Bias: Generative AI models can perpetuate or even amplify biases present in the training data [1]. This can lead to the generation of biased content.
3. The adoption of generative AI by businesses comes with a degree of ethical risk [6]. Organizations must ensure the responsible use of generative AI by ensuring it is accurate, safe, honest, empowering, and sustainable.
4. Data Privacy and IP Leakage: Generative AI systems can expose organizations to risks such as data privacy, IP leakage, and operational risk [7].
5. Lack of Verification: Without external verification or fact-checking, users might be misled by the information produced by generative AI [5].

6. **Citations:** Can't cite exact sources because they don't have access to the internet and don't remember where the training data came from. It may therefore be that sources are invented.

To mitigate these risks, it's important to use high-quality, diverse, and representative data for training, continually update the AI model with new data, always corroborate the outputs of generative AI with other trusted sources, and educate users about the strengths and limitations of the AI [5].

Status quo of the ccTOPP project

When dealing with problem statements, contradictions and the matrix, one can either think of an 'integrated' approach or a specific application of preselected Inventive Principles (IPs). 'Integrated' here means accord. to [xx], that an idea or a problem is described precisely to the LLM together with integral work assignment: to work out the improvening/worsening parameters, then to derive IPs and suggest solutions to overcome the contradiction by applying the principles. This is a rather complex and comprehensive workflow. And bringing light into each task for the LLM requires to first break it down into the constituting commands.

Based on this insight, the current focus within the ccTOPP project is to develop and explore prompting for the application of specific IPs. The above-mentioned prompting methods are used to investigate their impact and effectiveness.

We structured the approach in 3 steps (this is aligned with the approach in "TRIZ and Generative AI-V3.0"):

1) **What does the user need to know?**
   - A general understanding of the 40 Inventive principles
   - Optionally: how to define engineering/physical contradictions
   - Optionally: How to apply the contradiction matrix and derive suitable IPs
2) **What input does the user need to prepare for genAI input?**
   - A problem statement: a clear description of the engineering problem or the system where an improvement is desired
   - A detailed knowledge of the (sub-) IP to use
   - Further details explaining the principle, e.g. examples
3) **What can the user expect from the AI?**
   - Possible directions for improvement, worthy goals
   - A list of specific ideas to solve the problem

Additionally, the impact of the specific prompting/priming methods is investigated:

- What degree/amount of details of the IP is needed for a certain quality of results?
- What is the impact of delivering specific examples for the IP respectively sub-IP?

Summarizing the beforehand explanations, 4 levels of interactive prompts have been setup and applied (Fig. 2):

	Prompts			
	#1	#2	#3	#4
Input				
technical system to be improved	X	X	X	X
possible directions for improvement, worthy goals	X	X	X	X
inventive principle	X	X	X	X
components of the technical system		X	X	X
related sub-principles			X	X
at least 1 example for each sub-princple				X
Output				
ideas to improve the system (overall)	X			
ideas to improve each component of the system		X		
ideas for each sub-principle to improve the system (overall)			X	
ideas for each sub-principle to improve each component of the system				X

Fig. 2 Setup of the investigated prompts

As an example, we apply the 4 prompts to the task of improving an office desk. The prompts are setup non-interactive. This means, that the specific product to be improved (office desk) as well as possible directions for improvement & worthy goals are integrated into the prompt.

The prompts have been tested with Microsoft Copilot (based on ChatGPT 4) and ChatGPT 3.5. A comparison of the results in two directions is derived:

a) within an LLM, comparison of the prompt levels
b) between the LLM, same prompt level

In the following, the prompts #1 and #4 are shown exemplarily. If interested in all prompts, please contact the authors.

Non-Interactive Prompt #1 (no components, no sub-principles, no examples):
We want to apply TRIZ inventive principle 1 to improve a system.
The system is an office desk.
The inventive principle 1 is described the following way:
Principle 1: Segmentation
The directions for improvements are a) weight, b) reliability and c) costs.
Apply the principle to improve the system towards the intended direction and summarize the output in tabular form and give advantages and disadvantages for every idea you produce.
Summarize the output in tabular form.
The first column contains the intended direction for improvement.
The second column contains the specific idea for improvement.
The third column contains the advantages for each idea.
The fourth column contains the disadvantages for each idea.

Non-Interactive Prompt #4 (components, sub-principles, examples):
We want to apply TRIZ inventive principle 1 to improve a system.
The system is an office desk.
The inventive principle 1 is described the following way:
Principle 1: Segmentation

The sub principles and related examples are:

a. Divide an object into independent parts
Example of usage of sub-principle a: To enable IKEA to sell more furniture, the pieces of furniture have been broken down into independent parts that can be easily assembled according to a construction plan. This means you don't have to buy the whole wardrobe as one piece, as was previously the case.
b. Make a system easy to put together and take apart
Example of usage of sub-principle b: There are toothbrushes that have a brush head that can be detached from the head of the toothbrush. This has the advantage that you can replace the old head with a new one without replacing the handle.
c. Separate according to a condition or parameter
Example of usage of sub-principle c: There are tables and benches with retractable legs. This means that the legs were separated from the main part and made movable in comparison to the previously common designs. Depending on whether you want to use or transport them, one or the other is preferred.
d. Increase the degree of fragmentation or segmentation
Example of usage of sub-principle d: The finer the magnetic powder, the better the field lines of a magnetic field can be visualized.
e. Transition to microlevel
Example of usage of sub-principle e: If magnetic powder is ground so finely that the particles remain in suspension with oil, a ferrofluid is produced.
The principle is also known as fragmentation or division.

The sub-principles are normally used on components or elements of a system.###

The directions for improvements are a) weight, b) reliability and c) costs.

List the components of the system.

Apply all sub-principles a. to e. to all components, respectively: start with component 1 and apply all subprinciples to the component 1. Then continue with the next component and so forth.

Summarize the output in tabular form with the following format:

The first column contains the components of the system.

The second column contains – for each component – the applied sub-principles. Take one row for each pair "component/sub-principle".

The third column contains the intended direction for improvement.

The fourth column contains the specific idea for improvement.

The fifth column contains the advantages for each idea.

The sixth column contains the disadvantages for each idea.

Comparison and summary of the results:

The intended out format is (Fig. 3):

Component	Sub-Principle	Direction for Improvement	Specific Idea for Improvement	Advantages	Disadvantages

Fig. 3 Structure of requested output

A bird's-eye view (not to be read in detail!) on the results of MS Copilot and ChatGPT 3.5 looks like this (Fig. 4):

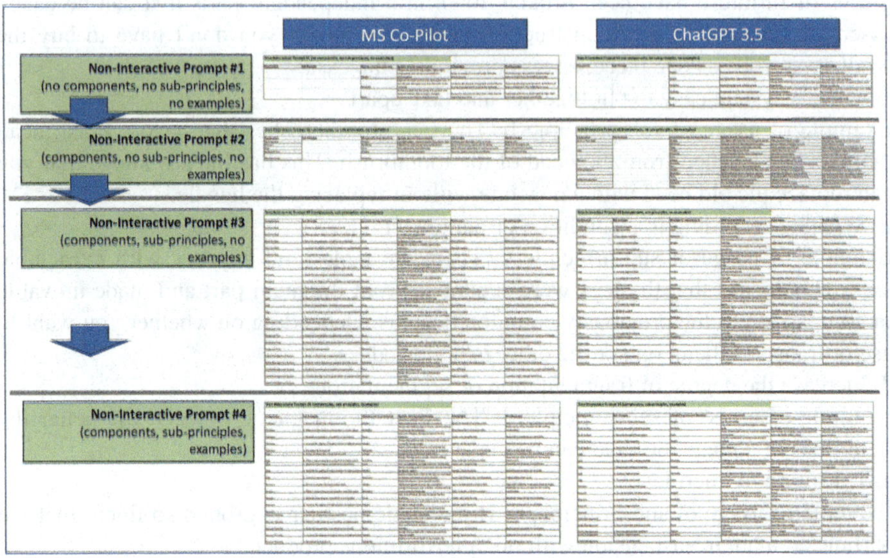

Fig. 4 Bird's-eye view on prompting output

The main take aways are summarized in the following:

- The more specific the input data are, the more specific the results are.
- If you want a specific output format, you have to specify this in advance.
- Without requesting to list the components of the system, the output is more general. Then the main categories (here) are the directions for improvement.
- By requesting to list the components, one idea per component is delivered. But the exact list of components is not reproduceable. If you intend to improve specific parts, you have to prime it.
- by priming the sub-principles and request the application of each sub-principle to each component, the quantity of output is increased (because it is requested).
- By additionally priming one example for each sub-principle, the output seems to get slightly more specific. But a clear increase in output quality cannot be seen.
- ChatGPT 3.5 seems to be more responsive and accurate in terms of following the instructions. The MS Co-Pilot sometimes does not execute the full request. E.g. instead of applying all sub-principles to all components respectively, the Co-Pilot just lists an extraction of that.
- The delivered ideas are not too detailed and a good starting point for further ideation, see the following example.

As an example, the output of ChatGPT 3.5 for the prompts #1 – #4 for improving the tabletop/ desk surface is shown. The output of prompt #1 has no focus on the component as for that we haven't requested this (Fig. 5).

	Sub-principle	Direction for Improvement	Idea for Improvement
Prompt #1	here no focus on specific components	Weight	Modular desk with detachable components
		Weight	Use lightweight, segmented materials
		Reliability	Interlocking segments without screws
		Reliability	Replaceable parts for easy maintenance
		Costs	Standardized, interchangeable segments
		Costs	Flat-pack design for shipping
Prompt #2		Weight	Use lightweight materials like honeycomb panels
		Reliability	Modular design for easy replacement parts
		Costs	Use less expensive, sustainable materials
Prompt #3	a. Divide into parts	Weight	Use modular panels that can be rearranged or replaced easily.
	b. Easy assembly	Costs	Design interlocking panels with no need for tools.
	c. Separate by condition	Reliability	Use materials that respond to heat/cold to prevent warping.
	d. Increase segmentation	Weight	Use a honeycomb core structure inside the tabletop.
	e. Transition to microlevel	Costs	Incorporate fine, lightweight, but strong mesh structures.
Prompt #4	a. Divide into independent parts	Weight	Design the surface in two or more sections that can be connected
	b. Easy to assemble and disassemble	Reliability	Use modular panels that click together without tools
	c. Separate according to condition	Costs	Make sections with different materials depending on the need
	d. Increase fragmentation	Weight	Create a honeycomb structure inside the desk surface
	e. Transition to microlevel	Costs	Use a composite of fine wood particles for surface manufacture

Fig. 5 Output of ChatGPT 3.5 for improving the tabletop of an office desk. The prompts #1 – #4 are explained before

In our opinion – beside the quantity of output – also the quality and the detailedness of the output is increased to some extent. But we had expected a more pronounced difference. As this article reports the first investigations for working with genAI and the IP #1 'segmentation', we hope soon to be able to further tailor the prompting in order to get more precise and sophisticated ideas

3 Conclusion

We investigated the impact of the priming prompts when requesting genAI generated ideas based on Inventive Principle #1 'Segmentation'. Delivering more details within the input like mentioning the sub-principles or adding specific examples increases the quantity and to some extend also the quality output. Further dependencies and design variables have to be investigated, e.g. …

- how important it is, that the given examples fit to the characteristics of the technical system.
- if more examples lead to a higher quality and higher specification of the generated ideas
- how to deal with the application of multiple Inventive Principles
- in general, if the effort in specifically priming the LLM is worth with respect to the obtained results.

References

1. Cangiano A., Ahuja R.: Generative KI: Grundlagen der Prompt-Technik. https://www.coursera.org
2. Coursera Staff: "What Is Prompt Engineering? Definition and Examples". https://www.coursera.org
3. Pheunghua T.: TRIZ and Generative AI (Example of Prompt). www.researchgate.net (2023)
4. Pheunghua T., Adunka R.: TRIZ and Generative AI-V3.0. www.researchgate.net (2024)
5. Dilmegani C.: 5 Risks of Generative AI & How to Mitigate Them in 2024. https://research.aimultiple.com/ (2024)
6. Baxter K., Schlesinger Y.: Managing the risks of generative AI. https://hbr.org/ (2024)
7. Shattuck S., Doria C.: Understanding Generative AI Risks: A Comprehensive Look at the Top 7 Generative AI Risks for Businesses. www.credo.ai (2023)

Author Index

O. Mayer (Hrsg.): TRIZ 2024, https://doi.org/10.1007/978-3-662-70698-5

Made in United States
North Haven, CT
18 October 2026

20546952R10091